Medical Mycology in the United States

A Historical Analysis (1894–1996)

MEDICAL MYCOLOGY IN THE UNITED STATES

A HISTORICAL ANALYSIS
(1894–1996)

by

Ana Victoria Espinel-Ingroff

Medical College of Virginia Campus of
Virginia Commonwealth University
Richmond, VA
USA

KLUWER ACADEMIC PUBLISHERS
DORDRECHT / BOSTON / LONDON

Library of Congress Cataloging-in-Publication Data is available.

ISBN 1-4020-1067-2

Published by Kluwer Academic Publishers BV,
PO Box 17, 3300 AA Dordrecht, The Netherlands.

Sold and distributed in North, Central and South America
by Kluwer Academic Publishers,
101 Philip Drive, Norwell, MA 02061, USA

In all other countries, sold and distributed
by Kluwer Academic Publishers, Distribution Center,
PO Box 322, 3300 AH Dordrecht, The Netherlands

Printed on acid-free paper

Printed and bound in Great Britain by Antony Rowe Limited

Table of contents

Foreword

In our contemporary world, scientific heritage is often forgotten. Many important contributions during the past 100 years in research, teaching, and diagnosis have had a profound impact upon the evolution of Medical Mycology in the United States.

This book, written by Dr Espinel-Ingroff, makes a significant contribution towards ensuring that those who have laid the foundation for our current Medical Mycology infrastructure are recognized for their scholarly and authoritative contributions. This book is truly a landmark publication towards ensuring that the past, present, and future are connected to each other.

Libero Ajello, Ph.D
Michael R. McGinnis, Ph.D

Acknowledgments

The invaluable editorial assistance and creative reinforcement of my husband, David Ingroff, was especially important during the preparation of this book. Special thanks go to Drs Libero Ajello, Michael McGinnis, and William Blake for their support, guidance, encouragement, and continuous editorial assistance. I would like to thank the numerous medical mycologists who responded to the questionnaire and/or made themselves available for interviews. Also, Drs Glenn Bulmer, Timothy Cleary, Arthur DiSalvo, Judith Domer, Norman Goodman, Morris Gordon, Carlyn Halde, Richard Hector, Dexter Howard, Leo Kaufman, George Kobayashi, Kyung Joo Kwon-Chung, Geffrey Land, Donald MacKenzie, Thomas Mitchell, John Rippon, Stanley Rosenthal, Ira Salkin, Wiley Schell, Margarita Silva-Hutner, Jim Sinski, Paul Szaniszlo, John P. Utz, and Mrs Norman Conant provided invaluable archival documents, personal papers, photographs and unpublished historical documents that gave greater depth to this study.

I owe a debt of appreciation to the following individuals for their assistance during the gathering of the data including the mailing of the questionnaires: Drs Michael Rinaldi and William Merz, President and Treasurer, respectively, of the Medical Mycological Society of the Americas in 1994; the Medical College of Virginia Tompkins McCaw's Library week-end staff and the Library Interloan Service; J. Kerr, Archivist of the American Society for Microbiology; Drs W. Dismukes and M. Saag and Ms C. Thomas of the NIAID, Mycoses Study Group; Dr Dan Sheehan and T. Webster of Pfizer Inc., USA, Pharmaceuticals and Mr. Robert Scott from Merck, US Human Health. Sincere appreciation is due to Ms Joan Peters for her secretarial assistance during the preparation of this manuscript.

Finally, special recognition is due Dr Arthur DiSalvo, whose persistent support was most valuable in accomplishing the publication of this book.

Preface

The genesis of this work began with my desire to determine how medical mycology has developed as a science in the United States. This idea evolved from my search for a topic for my doctoral dissertation. My original idea was to conduct historical research on the development of antifungal drugs, which has been the primary focus of my career for the last twenty years. However, when I attempted to obtain access to archival documents from pharmaceutical companies involved in the development of such drugs, both in Europe and the United States, their uniform response was that such archives were closed to individuals outside the company. At this point, although it appeared to be a formidable task, I decided to broaden my research and investigate the development of the discipline in this country. Next, I determined the extent that this investigation was needed and who would utilize the results. During the 1993 Focus on Fungal Infections meeting in Tucson, Arizona and the Conference on *Candida* and Candidiasis in Baltimore, Maryland, I conducted informal interviews with medical mycology leaders who were involved in patient care, research, and the training of future medical mycologists. I also interviewed retired leaders by phone. My initial findings were that most of these medical mycologists had a number of concerns regarding the current state and future development of the science in this country.

Since the beginning of the 1990s, the perception was that the discipline was changing rapidly and that training and research programs in medical mycology were decreasing in number and scope in the United States. On the other hand, a significant increase in the incidence and prevalence of nosocomial yeast and mould infections has been evident. As a result of improved management protocols, AIDS and other immunosuppressed patients live longer, but remain highly susceptible to life-threatening opportunistic fungal infections. Since it is not mandatory that physicians and other health care professionals report mycotic diseases, it is difficult to compare their incidence and prevalence with that of other infectious diseases. Leaders in the field felt that additional trained personnel were essential to respond to the increased incidence and prevalence of debilitating and fatal fungal diseases. In the area of research development, the concern was that medical mycology had not made the necessary transitions at the same pace as the other branches of microbiology in the 1970s, when research moved from the whole organism to its molecular biology and genetics. Medical mycologists also became uncertain regarding the sources of funding for training and research because of the significant decreases in these funds that began in the early 1980s. With these issues in mind, I conducted a historical inquiry into the development of the discipline, which provided insights into why and how its development and progress had been affected during the last 100 years.

What is the field of medical mycology? This discipline is concerned with the study of medically important fungi and fungal diseases in humans and other animals. With this description of medical mycology in mind, this book may serve as a reference tool or a source of learning about the discipline for individuals who are interested or involved in the different aspects of medical mycology: patient management; clinical, basic, and applied research; laboratory diagnosis; as well as education and training. It will also be a reference book for medical, microbiology and medical technology students. It provides overall and detailed chronological descriptions of the most significant scientific, educational, and technological medical mycologic events that took place from 1894 to 1996 in the United States.

What other publications could supplement the information provided in this book? A search of the literature revealed that there are no comprehensive historical studies of the development of medical mycology in the United States. Uncovered were a few monographs and a number of obituaries regarding pioneer American medical mycologists and works concerning certain epochs and topics in this country and Europe. Some of these publications are chronologies that are considered summary generalizations by historians. In addition, my interviews indicated that efforts were made in the 1980s to publish a biographical book regarding medical mycology's pioneers up to 1960. However, it appears that financial difficulties precluded publication of that work. Important information was obtained from several of these unpublished chapters and other historical publications cited in the list of references. In 1986, Ainsworth published his *Introduction to The History of Medical Mycology* and Drouhet published in 1992 a handbook on mycology that included an introductory chapter on the history of education and training in medical mycology worldwide. These works, however, do not provide an in-depth consideration of any particular country or area. Appendix D briefly summarizes these and other related publications.

Introduction

Fungal infections occur throughout the world, but some of them are more predominant in certain geographic areas. Although fungal diseases usually are not considered to be as common as bacterial and viral diseases, they are frequently associated with severe morbidity and mortality. Among the superficial fungal infections, tinea pedis (athlete's foot) and pityriasis versicolor are among the world's most common diseases.

The increased frequency and severity of mycotic diseases have prompted greater research efforts in the development of antifungal drugs. However, despite those efforts, in 1996 there were only six antifungal drugs licensed in the United States for use against systemic fungal infections. Furthermore, the newer candidate drugs required further evaluation of their efficacy because their spectrum of effect and application in the eradication of the various mycoses remained unanswered. The need for more effective, economical, less toxic, and easier to administer antifungal agents to treat systemic fungal infection was clearly evident. The evaluation of new compounds and management approaches demanded more training and research programs.

It never has been easy to acquire formal training in medical mycology. The training of a medical mycologist has been considered a twentieth century innovation. As the incidence of fungal diseases increases, there is clearly a need for more and better trained individuals. Two concerns require immediate attention: the replacement of the diminishing number of senior mycologists and the provision for broader based multidisciplinary training programs to better meet the new challenges of the 21st century. More training programs should be established at both the pre- and postdoctoral levels with molecular biology as a major component. These endeavors demand the provision of more financial support and allocation of other key resources to medical mycology in this country.

The largest portion of federal funding for health oriented research is administered through the National Institutes of Health (NIH). Kennedy's (1990) analysis of policy effects on federal support of health research revealed several negative regulatory factors. Although appropriations for NIH grew steadily during the decade of 1980 to 1990, expenditures for training and research contracts decreased. Direct NIH investments in training rose slightly, but fell dramatically when discounted for inflation. Expenditures for research related to AIDS accounted for approximately 20% of the overall extramural programs, while other training and research contracts declined. As governmental sources of funding are reduced, the academic and medical communities seek more support for educational and research activities from industry. These developments have led to closer collaborations between academic and corporate researchers. Due to varying economic conditions in the future, corporate funding can be expected to become even more restricted and difficult to obtain.

What has been the trend for funding of medical mycological training and research? In the beginning of this century, philanthropic and private organizations (e.g., the Harvard Traveling Fellowship, the College of Physicians, the Rockefeller Foundation, and the Brown-Hazen Foundation) provided grant support for training and research in medical mycology. However, those contributions were usually small and generally had ceased to exist by 1996. At that time, some fellowships and grants for medical mycology were being supported by Pfizer Inc., US Pharmaceuticals Group and Janssen Pharmaceutica.

The purpose of this study was to investigate, in a historical context, the development of medical mycology training and research in the United States. The development of this discipline has been assessed within the context of scientific progress, as demonstrated by the creativity and scholarly contributions from training, research, and technological activities directed toward the control of fungal diseases. The study covers the historically significant, as well as decisive educational, scientific and technological events that have guided the development of medical mycology in this country since the late 1800s. It includes key figures who led the field and the effect of their ideas and visions on the evolution of the discipline. However, the scope of this book does not include two important fungal pathogens, *Paracoccidioides brasiliensis* and *Pneumocystis carinii*. The former is important in those areas of Latin America where it is endemic, and the latter was considered phylogenetically closely related to fungi only in the late 1980s.

Narrative description was the first phase of this study and it involved the critical evaluation of documents to uncover significant facts. The narrative description was accompanied by an analysis of the causes of historical events that actually took place during those first 100 years, some reasons for related human behavior, and the perceived impact on the development of the discipline in the United States from the late 1880s into the 1990s.

The process of choosing what is considered significant in historical research is called the method of selection by historians. For the present study, the selection involved the filtering of the facts that explained the causation of the events relevant to what was important to the development of the discipline. It was performed according to factual data and the social value being assessed, which was defined as the control of fungal diseases.

This study analyzed changes in the foci of training and research, the composition of groups, the institutions that resulted (scientific societies, research groups), and past and present leaders. Both the uniqueness and the commonness of the events were investigated. The commonness allowed the grouping of related events or the categorization of the events by the identification of similar features. The major contributions of past and current medical mycology leaders also were evaluated by themselves (if alive) and others through the questionnaire. The grouping and the identification of unique characteristics suggested problems, helped categorize the data, and stimulated the formulation of conclusions. These analyses also helped to explain new directions in which the discipline was moving.

The use of concepts helps historians to avoid erroneous or oversimplified causative explanations. Two concepts, defined as the abstract ideas that refer either to a class of phenomenon or to certain characteristics that phenomena have in common, were used to generate data on the influence of an individual on the direction and career of others. One concept was the formal training connection, with special attention to the influence of major professors on students in their masters and doctoral programs in medical mycology. The other concept was either informal training and mentorship, or a combination of both, where the influence was not connected with a formal degree, but was regarded by the respondent as vital for his/her career and development as a mycologist. "Training Trees", Appendix B were then created to reflect the genealogy of the discipline. They provide a model to display the information gathered on the training and education of medical mycologists in the United States.

A number of data sources were examined for relationships between specific contributions and the knowledge derived and disseminated. For the scientific contribution sections, the data collection focused on selected published papers of significant scientific and technological events. Such sources are usually considered secondary sources in historical studies, but they are regarded in this book as direct records of the era in which they were published. More than 2000 scientific articles were examined, the selected ones are listed in the references. The criteria used to select the articles focused on first contributions such as: the first description of a fungal disease or fungal pathogen, pertinent milestone publications from outside the United States, the development of antifungal agents and their evaluation, the establishment of therapeutic regimens, the resolution of taxonomic issues, useful laboratory diagnostic tests, important technological inventions, trends or events that changed the direction of the discipline, and discoveries that led to meaningful future research breakthroughs. No reviews or books will be found among the references because only articles describing original discoveries were used as data sources. However, important medical mycology books were reviewed and an annotated list is included under Appendix C.

Other primary sources comprised materials such as the minutes and bulletins of the Medical Mycological Society of the Americas (MMSA), archival material from the American Society for Microbiologists (ASM), and the personal files of medical mycologists. The latter sources included correspondence pertaining to their research and training experiences and their curriculum vitaes. The first training and research programs in medical mycology originated at Columbia University and Duke University. The Duke University archival materials were not available, however, a training grant proposal describing this training and research program since its establishment was provided by Thomas Mitchell. Archival material from Columbia University that related to medical mycology was obtained from Margarita Silva-Hutner. The available unpublished chapters on the history of Medical Mycology at Tulane University and biographies of early leaders as well as numerous obituaries were examined and listed under References. The unpublished chapters are archival documents of the MMSA and were provided by Morris Gordon.

Oral history has been used as a primary source in other historical studies. This primary information source represents living resources or participants. Thirty-seven medical mycologists were carefully selected and interviewed to obtain their views and accounts of their work. This oral history helped to support other documents and provided useful in-depth perceptions of un-published information regarding the impact of training on the field.

Questionnaires are valid data sources in historical research and have been used previously to determine the degree of involvement in training and research activities. This study included an open-ended, infor-mal questionnaire that was answered by 100 of the 275 medical mycologists listed as members of MMSA living in this country and by 11 of the 36 physicians listed in 1993–94 as members of the National Institute of Allergy and Infectious Dis-eases, Mycoses Study Group (NIAID-MSG). The questionnaire provided invaluable information regarding the institutions where medical mycolo-gists have received formal or informal training ("Training Trees", Appendix B). It also contained questions concerning the availability of funding, the current state of the science, and the changes deemed necessary for further advancement of medical mycology. In addition, the perceptions of the respondents regarding important contributions and contributors to the development of the discipline in this country were provided (Tables 3 and 4). A summary of the analysis of the questionnaire is attached (Appendix A). The "Trees" were displayed during the 1995 and 1996 MMSA annual meetings and were reviewed and revised as necessary. How-ever, this genealogical listing should not be consid-ered exhaustive as it only reflects the information the author obtained from her efforts as noted.

I conceptualized the development of medical mycology into five periods, or "eras", according to changes in direction resulting from various scientific and educational contributions: (a) the "Era of Dis-covery", 1894 to 1919, (b) the "Formative Years", 1920 to 1949, (c) the "Advent of Antifungal and Immunosuppressive Therapies", 1950 to 1969, (d) the "Years of Expansion, 1970 to 1979 and (e) the "Era of Transition", 1980 to 1996. This book is structured around these eras as the following chap-ters discuss the origin and the development of medical mycology in the United States. The impor-tant milestone events that span these eras have been summarized in a table at the end of this Introduction under the title, "Developmental Eras in the History of Medical Mycology in the United States, 1894 to 1996".

Chapter I

Historically many fungi had been known before their role in animal and human diseases was dis-covered between 1835 and 1841. These early dis-coveries led to other important studies, both in Europe and Latin America, from that period into the early 1900s. Chapter I describes in detail those original contributions to medical mycology, or its "roots".

Chapter II

The "Era of Discovery" explores the key figures and their first descriptions of important mycotic diseases in the United States from the 1894 to 1919. The recognition of fungi as etiologic agents of systemic diseases by physicians, who were not trained as medical mycologists, marked the beginning of the discipline in this country. But, historical studies of key figures from related sciences during this period, considered the golden era of microbiology, have not acknowledged the value of these contributions to the study of infectious diseases. Even today the impor-tance of fungal diseases in public health is over-looked. The intent of this book is to assess the impact of those early medical mycology discoveries.

In 1987, Kass divided the history of the specialty of infectious diseases in this country into three broad phases. His first phase is the "colonial era" when "every physician was necessarily an infectious dis-ease specialist" (p. 745). The second phase encom-passes the years of the first discoveries of the microbial causes of disease. The third phase began with the advent of specific chemotherapeutic treat-ment after World War II, the era of the antibiotics. Kass' second phase closely corresponds to the origin of medical mycology in the United States. This book's "Era of Discovery" also corresponds with K.S. Warren's (1990) first developmental period for tropical medicine, which he considered a product of the colonization of the tropics.

Chapter III

This chapter describes the beginning of medical mycology training programs and the establishment of research laboratories during the formative years of the discipline from 1920 to 1949. Key pioneers during this era provided more logical classification systems that improved the scientific investigation of human fungal pathogens. Their epidemiologic studies facilitated the differentiation between infection and disease, thereby improving the understanding of the natural history of systemic mycoses. Beginning in the late 1920s, continuing throughout the 1930s, World War II, and the late 1940s, leaders in the field trained the second generation of medical mycologists and ensured that medical mycology continued its development in this country as a distinctive field of medicine.

Chapter IV

Chapter IV deals with the third developmental era, the "Advent of Antifungal and Immunosuppressive Therapies", which began with the discovery of nystatin in 1950 and amphotericin B several years later. This era corresponds with Kass' (1987) third phase in the development of the specialty of infectious diseases. That phase began with the advent of specific chemotherapeutic treatments after World War II, the era of antibiotics. However, the "Advent of Antifungal and Immunosuppressive Therapies" is a more complex period in medical mycology. This period encompasses the initiation of renal transplantation in conjunction with the use of immunosuppressive therapy for oncology patients and to avoid organ rejection.

Chapter V

During the 1970s, the "Era of Expansion", the direction of medical mycology's development changed as a consequence of the increased incidence of mycoses in immunosuppressed patients and important advances in technology. With the commitment of sizable new revenue sources, the discipline saw new centers and programs established, as well as a major emphasis in educating and training the "third generation" of molecular-based medical mycologists. In terms of commitment of resources and the number of scientists dedicated to the field, this era is considered the high water mark period to date for medical mycology in the United States. Built on the 1950s and 1960s, rapid advances in transportation and communication occurred during the 1970s, as well as public acceptance in utilizing them, a trend that escalated in the 1980s and 1990s. The latest technological discoveries and the experts behind them became more accessible. All of these events directly impacted the discipline.

Chapter VI

This chapter deals with the "Era of Transition" between 1980 and 1996 when the much higher incidence of human mycoses, because of the AIDS pandemic, again brought major changes to the discipline. During this era, research in molecular biology, cellular immunity, and genetics was either initiated or intensified and DNA-based diagnostic tests and epidemiologic tools were developed. Recent developments within the discipline of microbiology have often resulted from discoveries using molecular tools and the knowledge that DNA is a repository of genetic information. A large number of microbiologists as well as medical mycologists reoriented their research direction from the microbiologic aspects of fungal molecular models to their cellular functions. In 1986, Beeson compared these scientific developments to the "golden age of microbiology" in the late 1800s when bacteria were recognized as etiologic agents of disease. During that era, the fundamental phenomena of immunity also were recognized. The invention of the microscope played an important role in the earlier development of these microbiologic disciplines, but has had limited importance in contemporary times. The impact of these advances and how medical mycology has kept abreast with technology, as have related sciences, is accessed in this chapter. Another significant trend of this era was the retirement and death of numerous leaders in the field. Coupled with the major reductions in governmental and foundation funds, the result was a major shift in leadership and a much stronger emphasis on collaborative research.

Chapter VII

This chapter summarizes the five developmental eras of medical mycology in the United States and provides specific conclusions drawn from this historical study.

Developmental eras in the history of medical mycology in the United States, 1894–1996

Era	Major events
The Era of Discovery (1894–1919)	Discovery of the dimorphic fungi and establishment of their dimorphic nature Recognition of fungi as etiologic agents of disease including systemic diseases
The Formative Years (1920–1949)	Establishment of the first medical mycology training and research center at Columbia University (1926) and other higher education institutions and federal agencies Development of laboratory diagnostic tests and classification systems First epidemiologic studies led to the recognition that fungal diseases were prevalent First ecological studies
The Advent of Antifungal and Immunosuppressive Therapies (1950–1969)	Discovery of first antifungal agents (e.g., griseofulvin, nystatin and amphotericin B) Establishment of the relationship of severe opportunistic fungal infections with antibiotics (1950s) and immunosuppressive therapies (1960s) Awareness that fungal diseases cause morbidity and mortality among oncology and transplantation patients Development of rapid diagnostic tests, e.g., immunodiagnostic tests for coccidioidomycosis and histoplasmosis capsulati and the detection of fungi in tissues by fluorescent antibodies and selective fungal stains Transition from biological to cellular level studies of fungi, e.g., antigenic composition of *Cryptococcus neoformans*, dimorphism mechanisms, virulence factors, host-parasite interactions Founding of The Medical Mycology Society of the Americas (MMSA, 1961) Creation of new training and research centers
The Years of Expansion (1970–1979)	Expansion and improvement of laboratory diagnosis, e.g., commercial yeast identification systems, detection of fungal antigens and metabolites, exoantigen test development and standardization of methodology Establishment of the regimen of choice for the treatment of cryptococcal meningitis Expansion of basic research studies including the description of the perfect states of certain pathogenic fungi Increased awareness of the importance of systemic mycoses led to the targeting of Brown–Hazen grant support for new or existing medical mycology training and research centers and an overall expansion of training, e.g., postdoctoral, workshops, short courses, etc. Formation of the Mycoses Study Group (MSG)
The Era of Transition (1980–1996)	Increased incidence and prevalence of fungal infections among AIDS patients and larger numbers of oncology and transplantation patients Era marked by major transitions and conflicting events in the field: (a) Depletion and termination of federal and foundation support led to a crisis in training and the decline and closing of established training programs (b) Leaders retired or died but were not replaced at their institutions (c) Researchers became more specialized, focusing on specific aspects of fungal diseases (d) Leadership transition to new and collaborative groups formed with scientists from other medical disciplines (e) Initiation and expansion of basic research in molecular biology and genetics and expansion of other studies funded mostly by NIH (f) Application of DNA-based methodologies resulted in accurate and rapid diagnostic laboratory tests and reliable epidemiologic tools (g) Provision of standards for antifungal susceptibility testing and establishment of better therapeutic regimens for certain systemic diseases in patients with or without AIDS supported by pharmaceutical and corporate funds

Adapted from Espinel-Ingroff, Clin Microbiol Rev 1996; 9: 235–272.

I. The origin of mycology

Figure 2. David Gruby (1810–1898) (pp. 5–7) (From: Rosenthal, T. 1932. *Annals of Medical History, 4*)

Figure 1. Johann Lucas Schoenlein (1793–1864) (p. 5). (From: MMSA archives)

Figure 3. Raimond Jacques Adrien Sabouraud (pp. 7–8) (Courtesy of Dr. Rosenthal)

Figure 4. (above) Alejandro Posada (p. 8) (Courtesy of Drs. Sordelli and Gonzalez)

Figure 5. (right) Lewis David von Schweinitz (p. 9) (From: A. Wolf & T Wolf. 1947. *The fungi* New York: Wiley & Sons)

Etiologic agents: cause of disease

An important concept in the understanding of medical mycology is the causation of infectious fungal diseases. Evans (1991) stated that the knowledge of the causation of any infectious disease depends on both conceptual and technical aspects. The conceptual aspect is concerned with the knowledge of the natural history of the disease and its pathogenesis. The pathogenicity of microorganisms was not widely accepted until the mid-nineteenth century. The main reason for this disbelief was that, up to then, most people believed in spontaneous generation and that new organisms originated from inert material. Even Louis Pasteur was occasionally uncertain if the unintended organisms that occurred in his cultures were due to spontaneous generation. In 1861, Pasteur published his monumental study on spontaneous generation in which he clearly demonstrated that ordinary airborne dust contained microorganisms (Carter, 1991). Thus, the door was opened for the study of microorganisms as etiologic agents of infectious diseases.

The technical aspects of understanding infectious diseases depend upon advances in technology and the laboratory techniques available to visualize, grow, and identify the causative agents, duplicate and study the natural history of the disease in experimental animals, and describe how the etiologic microorganisms cause a specific disease. The invention of the microscope marked the beginning of the development of all of the disciplines that deal with microorganisms, such as tropical medicine (K.S. Warren, 1990), the specialty of infectious diseases (Kass, 1987), and medical mycology. The light microscope was developed at the end of the sixteenth century. Its creation has been attributed to Antonie van Leeuwenhoek (Evans, 1991; Fuchs, 1984; Smit and Heniger, 1975). In 1675, Leeuwenhoek, a lens maker, used his rudimentary microscopes to examine a variety of samples and described a number of microorganisms such as the protozoa. A year later, Leeuwenhoek published his description of bacteria, which could be considered the beginning of descriptive bacteriology. This date has traditionally marked the birthday of bacteriology.

Although Leeuwenhoek observed bacteria and other microorganisms, he did not associate them with the etiology of disease. Subsequently, others applied microscopy to the discovery of the etiologic agents of infectious diseases. It is widely recognized today that infectious agents had been and still remain as major causes of morbidity and mortality world wide (Kass and Hayes, 1988). The discovery of the etiologic agents of infectious diseases had a vital role in the development of numerous disciplines concerned with microorganisms and the infectious diseases that they cause, such as the mycoses. It is noteworthy that the first association of a microorganism, a fungus, with a silkworm disease was reported by Agostino Bassi in 1835 (Kyle and Shampo, 1979).

The origin of general mycology: European roots

The hyphomycetes: first discoveries

In his extensive historical review regarding the development of the hyphomycete taxonomy, Subramanian (1983) emphasized the importance of historical research for scientific development with a quotation from A.H.R. Buller, "I have long been of the opinion that no man can have a full conception of any subject upon which he may specialize until he has studied the history of its development". Although the study of fungal diseases in humans began in 1837, the contributions of early general mycologists to the understanding of fungi associated with mycoses must be acknowledged. Many fungi were studied in the mid-1800s prior to understanding their role in causing lower animal and human diseases. Among the estimated 1.5 million species of fungi that are known or believed to occur in soil, water, or associated with plants and animals (Hawksworth, 1991), only approximately 150 of them have been adequately documented as etiological agents of disease in humans. In contrast, more than 1200 species of fungi have been associated with plant diseases. It would appear, therefore, that the history of mycology began with the discovery of the numerous species of filamentous fungi (moulds), known as hyphomycetes, which include the filamentous fungi of the form-class Fungi Imperfecti. Most of the fungi isolated in clinical laboratories belong to the class Hyphomycetes. Mycologists such as de Bary, Saccardo, Mason, and Bessey have had a profound impact upon the development of medical mycology.

The Florentine, Pier Antonio Micheli, was probably the first investigator to give scientific names to some of the hyphomycetous fungi. He was a botanist for Cosimo d'Medici III, and later a botanist in the Botanical Gardens at Florence and Pisa. His publication in 1729, *Nova Plantarum Genera,* included drawings of his newly discovered fungi with notes of his discoveries. Although Micheli did his studies using a primitive microscope in comparison to the microscopes available today, he accurately compiled morphologic data that can be used today to recognize the fungi that he studied. His collection, consisting of 16 volumes of plates containing his drawings and a box of fungal specimens, is kept at the Istituto Botanico in Florence. Micheli's personal copy of his publication and his original drawings, from which the plates were reproduced, are now in the British museum in London (Subramanian, 1983). Among the genera that he described was the genus *Aspergillus*, and at least two of its species, *A. flavus* and *A. fumigatus*, are important agents of opportunistic infections in humans. Micheli was the founder of mycology, as well as one of the creative minds of Italy.

Micheli's great contributions had essentially no impact for nearly 100 years. In 1790, Hainrich Julius Tode published his *Fungi Mecklenburgenses Selecti*, one of the oldest descriptions of local mycobiota (fungal flora). This work included drawings of 120 fungal species which he maintained in his collection. Unfortunately, his collection of fungal specimens was discarded after his death (Subramanian, 1983). Between 1796 and 1801 the physician, Christian Hendrik Persoon, provided the foundation on which the modern classification of the fungi was developed by subsequent mycologists. Persoon's skill with a hand held microscope and analytic ability resulted in his classic publication, *Synopsis Methodica Fungorum*, that was published in 1801. This first comprehensive investigation of fungal taxonomy incorporated taxonomic studies of earlier workers, including the 15 genera described by Micheli. Medically important fungi described during the 1700s included members of the genera *Candida* (as *Monilia*), *Aspergillus*, and *Torula*.

Two additional major contributions were the tomes, *Observations I* (1809) and *II* (1815), by the German physician and botanist, Heinrich Friedrich Link (Subramanian, 1983). Link established several genera of the hyphomycetes that are important to medical mycologists. Among those are the genera *Acremonium, Cladosporium, Fusarium, Geotrichum,* and *Penicillium*. His collection of 16 000 fungal specimens was deposited in Berlin. In 1817, C.G.

Nees proposed nine new genera of hyphomycetes including *Alternaria* and *Verticillium*, which are two groups of fungi occasionally seen as laboratory contaminants; isolates of some species classified in the former genus are being recognized as important emergent pathogens in the immunocompromised host. In the same year, Gottfried Ehrenberg coined the term hyphomycetes to designate this group of fungi. Ehrenberg's greatest contribution was the discovery of the sexual process in the *Syzygites* (Mucorales). His division of fungi as either idiotocous (homothallic) or coenotocous (heterothallic) was the forerunner of understanding heterothallism that was demonstrated so meticulously by the experimental work of Albert F. Blackeslee in 1904 (Subramanian, 1983).

The years of synthesis

By 1824 several early European mycologists had described many of the more common hyphomycetes. Their contributions to mycology involved the systematic syntheses of the new mycological findings with the earlier discoveries. The Swedish mycologist, Elias Magnus Fries, dealt with the taxonomy of the hyphomycetes in the second volume of his *Systema Mycologyicum* between 1821 and 1822. Both Rogers (1977a,b) and Subramanian (1983) discussed the concept of mentorship or leadership influence. Fries' work was based on the contributions of Persoon, which were very influential on one of the first American mycologists, L.D. deSchweinitz, and hence upon early American mycology (Rogers, 1977a).

Fries initiated a new direction for the development of mycology that culminated with the exploratory and systematic works of two Italian botanists: Corda's *Icones Fungorum Hucusque Cognitorum*, that was published between 1837 and 1854 (Subramanian, 1983), and Saccardo's Volumes II and IV of the *Sylloge Fungorum* published in 1886 and 1906 (Saccardo, 1886, 1906). These two works synthesized the voluminous and scattered descriptive studies of earlier pioneers who had dealt with the hyphomycetes. Pier Andrea Saccardo proposed approximately 50 new genera and developed a simple artificial system for fungal classification based upon their appearance in their hosts. This classification relied on easily evaluated morphological characteristics of conidia (color, septation, shape, arrangement) and their conidiophores (single, united into sporodochia or synnemata). Saccardo's system of classification became the universally accepted one until a clearer understanding of conidiogenesis was achieved (Ziegler-Bohme and Gemeinhardt, 1990). The need was perceived later to develop a classification system that reflected phylogenetic relationships typified by the concepts behind anamorph, teleomorph, and holomorph (M. McGinnis, personal communication, July 22, 1993).

The origin of medical mycology

Even though the history of medical mycology involving humans began with Robert Remak, Johann Lucas Schoenlein, and David Gruby, it is important to remember that it was the Italian lawyer and farmer, Agostino Bassi, who in 1835 discovered the mycotic nature of the epidemic disease of the silkworms called muscardine. Bassi correctly concluded that because he had failed to produce the disease's whitish "efflorescence" that was seen on dead silkworms with chemicals, the disease was caused by "an extraneous parasitic fungus" (Kyle and Shampo, 1979, p. 1584). Bassi presented his data and conclusions to the faculty of the University of Pavia and his manuscript *Del mal del segno calcinaccio o moscardino* was published in 1835. Bassi's fungus was originally named *Botrytis bassiani* in 1835 by Balsamo-Crivelli, and subsequently renamed *Beauveria bassiana* in 1911 by Jean Paul Vuillemin (Ajello, 1975). Bassi's association of an animal disease with a fungus stimulated an interest in the investigation of human diseases that might be caused by other fungi. This led to the discovery in humans of the first fungal pathogen causing a form of tinea known as favus or tinea favosa.

The mycotic nature of favus was discovered almost simultaneously in the European countries of Germany, Switzerland, and France (Rosenthal, 1932). In 1837, the Polish-born physician, Robert Remak, observed the presence of spherical bodies (arthroconidia) and ramified fibers (hyphae) in favus crusts on the scalp of infected individuals. Remak failed to associate these elements with the infection because he did not realize he had seen a fungus until he read Johann Lucas Schoenlein's paper (1839) regarding the "vegetable" (fungal) nature of favus. Since other investigators found similar mycelial elements in clinical specimens from other diseases, the importance of the observations made by Remak

and Schoenlein remained obscure and ignored until David Gruby clearly demonstrated that tinea favosa was caused by a fungus during the presentation of his paper to the Academy of Science in 1841.

Medical mycology: European roots

Gruby's discoveries

S.J. Zakon and Tibor Benedek (1944) translated Gruby's important first six papers to commemorate the first 100 years of medical mycology. In 1841, David Gruby presented to the Academy of Science in Paris his first paper entitled, *Sur une vegetation qui constitue la vraie teigne* (Rosenthal, 1932; Zakon and Benedek, 1944). In this contribution, Gruby described the causative fungus of tinea favosa, and established the contagiousness of this disease. Gruby, who lacked the benefit of having a modern microscope, precisely described the "septate, small, transparent and colorless filaments or cylinders [hyphae] and the oblong or round corpuscles" (arthroconidia) in specimens from cases of tinea favosa (Zakon and Benedek, 1944, p. 158).

Gruby clearly attributed the contagiousness of the disease to the presence of the fungal elements that he had seen in all of the cases of tinea that he studied. He supplemented his observations in subsequent publications with the results of his experiments, which included the inoculation of a fellow professor and himself with material from a patient's lesions. Although his attempts to infect humans produced a little inflammation and slight suppuration of the skin, he was not satisfied with the results because they did not accurately reflect actual lesions. His inoculation of plants, birds, reptiles and mammals also did not result in infections (Ziegler-Bohme and Gemeinhardt, 1990).

Gruby did not name the etiologic agent of favus, it was accomplished later by Remak (Ainsworth, 1986). Remak had allowed Xaber Hube to publish his first results in Hube's doctoral thesis, *De morbo/ scrofuloso* (Kirsh, 1954). Later, Remak was able to successfully inoculate his own left forearm and to grow the fungus on apple slices. In 1845, he reported his investigations and named the etiologic agent of favus *Achorion schoenleinii* (now *Trichophyton schoenleinii*) in honor of Schoenlein in whose laboratory (University of Berlin, Germany) Remak was working at that time. Although Remak was the first investigator to observe the presence of fungal elements in a lesion of tinea favosa, Gruby's publications had a greater impact among his contemporaries. Modern historians consider that the development of medical mycology began with Bassi, Gruby and Remak (Ainsworth, 1986; Ajello, 1975).

Gruby expanded his investigations to the study of thrush in children. He was able to demonstrate in 1842 that the membrane seen in this oral disease was not an inflammatory exudate, but almost always a pure culture of what is known now as the yeast *Candida albicans*. Gruby conducted a mycological investigation of this organism and placed it in the genus *Sporotrichum*. A well documented description of *C. albicans* was provided by another Swedish physician, F.T. Berg (1846), one of Gruby's students, who had focused on the study of infant thrush. As with the etiologic agent of favus, *Candida albicans* was named by other investigators. Charles Robin renamed this fungus *Oidium albicans* in 1853 and W. Zopf (1890) introduced the name *Monilia albicans* (Benham, 1931). In 1923, this yeast was classified in the genus *Candida* by C.M. Berkhout. *C. albicans* had been discovered in 1839 by B. Langenbeck from Germany in the white patches of thrush in the oropharyngeal areas of a patient with typhoid fever. However, he thought it was the etiologic agent of typhus.

Gruby pursued the study of hair infections and published three additional papers within three years. The first of these, *Sur une espece de mentagre contagieuse resultant du developpement d'un nouveau cryptogame dans la racine des poils de la barbe de l' homme*, summarized the results of his investigation of fungal infections of the chin (tinea barbae). The etiologic agent of tinea barbae was named *Microsporum mentagrophytes* in 1853 by Robin. Gruby also recognized and distinguished both the ectothrix (*Trichophyton mentagrophytes*) and endothrix (*T. tonsurans*) forms of hair invasion.

In his biography of Gruby, Rosenthal (1932) described him as "an expert microbiologist but only a mediocre dermatologist" (p. 342). For Rosenthal, Gruby's papers provided good descriptions of the fungi, but his clinical descriptions of the infections were too succinct and abbreviated. They did not provide enough information that would allow others to adequately recognize the clinical entities he described. Despite Gruby's notable findings, his contributions did not clarify the prevalent confusion among physicians of his time concerning these infectious diseases. Those who accepted Gruby's observations tried unsuccessfully to find fungal elements in every infection of the skin, while his detractors were delighted with these failures and denied the authenticity of his observations. The advancement of a field depends, in part, upon the degree that its leading investigators are willing to take a risk and express their views.

Gruby's next paper (Rosenthal, 1932) on ringworm of the scalp (tinea capitis) added more confusion at that time to the understanding of this infection in England and France. The English school of thought accepted Willan's and Bateman's descriptions, whereas French physicians accepted the diagnostic descriptions of two laymen, the Mahon brothers. Thus, multiple clinical names were given to this type of tinea which added to the confusion. Fortunately, Cazenave (Rosenthal, 1932) in 1840 provided an adequate description of tinea capitis and differentiated it from alopecia areata, a baldness of non-parasitic nature. Gruby's paper on ringworm of the scalp was contemporary with Cazenave's paper. However, Gruby's clinical description of the infection was inaccurate, which resulted in the disease being given the wrong name. He also failed to mention that his patients were children. This time, Gruby named the pathogenic agent of ringworm of the scalp in children *Microsporum audouini* (now *Microsporum audouinii*) after the celebrated zoologist, Jean Victor Audouin (Rosenthal, 1932). Geoffrey Ainsworth (1986) believed that the name of this species is the oldest name still currently used for a medically important pathogenic fungus. Its only change was the required latinization of the name.

Gruby's last paper was entitled *Recherches sur les cryptogames qui constituent la maladie contagieuse du cuir chevelu d'ecrite sous le nom de teigne tondante (Mahon), herpes tonsurans (Cazenave)* (Rosenthal, 1932). In this paper, Gruby described *Trichophyton tonsurans*. This publication is considered the best of his six papers (Rosenthal, 1932). However, since it was the Swedish scientist, H. Malmsten who named *T. tonsurans* in 1845, Malmsten's publication is more widely known. Gruby discontinued his research studies after this paper and dedicated himself to patient care.

The concept of milestones to mark important and significant events in the development of a person or a society was adopted by Ajello (1975) as a means to highlight important contributions to our scientific knowledge to better understand diseases caused by pathogenic fungi. Ajello (1975) considers Bassi's contribution the first milestone in the development of medical mycology, because he was the first person to prove that microorganisms could cause a disease in animals. Bassi was not only the first "medical mycologist", but also the first individual who demonstrated that a microorganism, a fungus, could cause an infectious disease. The second milestone was attributed by Ajello (1975) to both Remak and Gruby. These two individuals first showed the etiologic role of fungi in human disease. Ajello (1975) states that "Gruby's first three papers accurately and succinctly described the clinical expression and in vivo microscopic details of favic crusts and the tissue form of *T. schoenleinii*" (p. 5).

Other early European contributors

Otto Busse (1894) at Virchow Hospital and A. Buschke (1895) in Zurich studied and reported separately the first *Cryptococcus neoformans* (as *Saccharomyces hominis)* case. The organism was first seen in a surgical specimen from a woman who had an abscess on the left tibia that was diagnosed as a sarcoma. They described the yeasts that they saw in tissue and culture as doubly contoured, spherical to ovoid cells. Busse was able to produce similar lesions in mice to those found in the patient's tissues and obtained a pure culture using a prune juice medium (Schwarz and Baum, 1964). Subsequently, the patient developed lesions on the face and other areas of the body and died. Busse reported his work on July 7, 1894 to a local medical society. Buschke also attempted to culture the organism, but Busse's investigation was conducted previously. The pathologist Busse was born in Germany and was a student of Paul Grawitz, an assistant of Rudolph Virchow. Other European investigators studied the natural history of this organism including Francesco Sanfelice (1894) who gave it the name *Saccharomyces neoformans,* Jean Paul Vuillemin (1901) who transferred it to the genus *Cryptococcus,* and D. von Hansemann (1905) who described the first case of cryptococcal meningitis.

The reports of the role of *Aspergillus* spp. in human pulmonary diseases by the German physician, Rudolph Virchow in 1856 (Ziegler-Bohme and Gemeinhardt, 1990) at the University of Berlin, Germany, and the work of Aldo Castellani on the use of biological criteria for the identification of pathogenic yeasts (Binazzi, 1991; Seeliger and Seefried, 1989) are additional important European contributions to the development of medical mycology in the early years.

The next important European contribution belongs to the German, Paul Grawitz and the Frenchman, Emile Duclaux, who simultaneously in 1886 grew several pathogenic fungi in pure culture (Ajello, 1975). The contamination of cultures by the fast growing species of *Aspergillus* or *Penicillium* created the misconception that either all dermatophytoses were caused by a single fungus, or that the dermatophytic fungi were mutants of these fast growing fungi. The introduction of pure culture techniques was vital for resolving these erroneous ideas regarding the etiologic agents of the dermatophytoses and of other fungal diseases. In 1877, George Thin (King, D.F. and King, L.A.C., 1988) used a pure culture technique and a variety of liquid media and "meat-gelatin" to prove that the pathogenic fungus *T. tonsurans* "is not related to ordinary fungi... with which it had been up to that time confounded" (p. 547). These and other early mycologists, who worked from 1841 to the end of the century, prepared the way for Raimond Sabouraud who decisively established the plurality of the etiologic agents of ringworm infections (Ainsworth, 1961).

Sabouraud's contributions

Raimond Jacques Adrien Sabouraud marked the period of transition from the dermatophytoses (the now overshadowed infections) to the systemic fungal diseases (the then overshadowed infections). Ajello (1975) aptly wrote that "the one man who truly revolutionized our concepts of the dermatophytes and who immeasurably contributed to the development of medical mycology was Raimond Sabouraud" (p. 6). Sabouraud represents Ajello's (1975) fourth milestone. Sabouraud was able to crystallize and organize the accumulation of fifty years of scattered observations regarding the role of pathogenic fungi in ringworm infections. He took advantage of the newly developed pure-culture technique and integrated the mycological and clinical aspects into a comprehensive concept. His contributions are contained in his publication *Les Teignes* (Sabouraud, 1910), which was the most scholarly contribution of his time regarding the dermatophytes, as well as an invaluable contemporary reference tool.

Although many of the common dermatophytes were described by 1910, it was Sabouraud who classified the dermatophytes into the four genera: *Achorion*, *Epidermophyton*, *Microsporum*, and *Trichophyton*. He developed media for their isolation, identification, and maintenance and then established the taxonomic criteria for their identification. He proposed the name *Epidermophyton inguinale* for the fungus (now called *E. floccosum*) that causes eczema marginatum e.g., tinea cruris, ringworm of the groin. Sabouraud and Noise introduced the use of X-rays as a therapeutic procedure to reduce the prevalence and incidence of dermatophytosis (tinea capitis) in Europe. This procedure also alleviated an important socio-economic problem. The length of hospitalization to cure these infections was reduced from 2 years to 3 months and the segregated schools required for the education of affected children with ringworm were no longer needed.

Early Latin-American contributors

In 1892, the Argentinean soldier Domingo Escurra was referred to Alejandro Posada, a student of Robert Wernicke, professor of pathology at the University of Buenos Aires, for confirmation of a possible diagnosis of cutaneous tuberculosis. Verrucous lesions and lymphadenopathy had developed in this patient following a presumed bite by a spider on his cheek. Posada (1892) found the characteristic spherules of *Coccidioides immitis* in the patient's tissue sections but he thought this organism was the etiologic agent of "mycosis fungoides". Escurra died as a result of this fungal disease in 1889 and his head was preserved and remains as a specimen in the pathology museum of the University Hospital in Buenos Aires. Four years after Posada's discovery, Emmet Rixford and Thomas C. Gilchrist (1896) compared the organism that they found in tissues from a farm laborer living in the San Joaquin Valley of California with Posada's accurate descriptions of the spherule cells and concluded that they were dealing with the same organism (Chapter II).

In 1900, Guillermo Seeber, a student of Wernicke, reported the first case of rhinosporidiosis. The patient was a 19-year-old farmer who had a large nasal lesion that impeded breathing. Seeber (1900) believed that this disease was caused by an organism related to that of Posada's case. In 1923, Ashworth in Scotland definitely concluded that the organism isolated by Seeber was a fungus and named it *Rhinosporidium seeberi*.

At the Instituto Oswald Cruz that was founded in 1900 at Rio de Janeiro, Brazil, Paulo P. Horta and other investigators discovered black piedra in 1911. The clinical aspects of this infection had been described in 1876 by Nicolas Ozorio and Posada Arango in Colombia (Drouhet, 1992). The etiologic agent of black piedra was described as *Trichosporon hortai* in 1913 by Emile Brumpt from France and reclassified as *Piedraia hortae* (as *hortai*) by Olympo da Fonseca and de Arêa Leão (1928) in Brazil to honor Horta. da Fonseca, originally a parasitologist, was trained in the United States as a Fellow of the Rockefeller Foundation under the direction of the general mycologists Roland Thaxter (Harvard University), Charles Thom, and the dermatologist Caspar Gilchrist, who described the first case of blastomycosis. Da Fonseca also visited the laboratories of Raimond Sabouraud, Maurice Langeron, A. Guilliermond and other important European scientists. Tinea nigra was described in 1898 in Colombia by Montoya and Flores. In 1891, Alexandre Cerqueira had provided a well documented description of this disease; however, this report did not appear in the literature until 1916 when Cerqueira's son reported the case. The etiologic agent of this disease was isolated by Horta (1921) and he named the organism *Cladosporium werneckii* (now *Phaeoannellomyces werneckii*). In 1908, Adolfo Lütz was the first to recognize that paracoccidioidomycosis was a fungal disease, while he was the director of the Bacteriological Institute of Sao Paulo. Lütz obtained his degree in Switzerland. The role of phaeoid fungi as etiologic agents of chomoblastomycosis was established by numerous case reports from Latin America, including the four cases of *Fonsecaea pedrosoi* (as *Phialophora verrucosa*) reported from Brazil by A. Pedroso and J.M. Gomes (1920) following the case described by M. Rudolph (1914). The fungus was named *Hormodendrum pedrosoi* by Brumpt in 1922, because he thought it was different from Medlar's (1915) and Lane's (1915) isolate (*Phialophora verrucosa*) (Chapter II). In Argentina, Pablo Negroni (1936) proposed the new genus *Fonsecaea* to accomodate isolates that produce *Cladosporium*-like and sympodial forms of conidiogenous cell formation.

The origin of general mycology: United States

Lewis David von Schweinitz, who is considered the first American mycologist, is lavishly praised as the "mycological father of us all ... of North American mycology" by the mycologic historian Donald Rogers (1977a). von Schweinitz was educated in a German seminary, became a priest of the Moravian church in 1801, and obtained his Ph.D. in 1817. During his years in Germany, his mentor was J. von Albertini, a dedicated botanist and mycologist. von Schweinitz started his studies of plants and fungi when he returned to the United States after receiving an appointment as administrator of the Church Estates in Salem, North Carolina. The list of his collection of over 1300 fungal species was completed in 1818 and it included elaborate descriptions of many new taxa. Published in 1822 under the title, *Synopsis fungorum carolinae superioris* (Rogers, 1977a), his work on American fungi predated Elias Magnus Fries' *Systema Mycologicum* in 1821. Although other mycologic works were published before von Schweinitz's four books, Rogers (1977a) believes that von Schweinitz was the first American mycologist. Earlier mycologic publications contain either references to probable fungal species without a name, or a description of the species, or only inadequate descriptions (Rogers, 1977a). They were at best mere catalogues or lists of fungi.

Many other botanically oriented mycologists have contributed to the development of mycology in the United States. Robert M. Page (Emerson and Humber, 1970), Lexemuel R. Hesler (Petersen, 1978), and Lee Bonar (Tavares, 1979) made important contributions through their leadership, teaching, and the direction given to graduate and undergraduate students. However, they also deserve recognition for their scientific contributions to the understanding of the nomenclature and taxonomy of pathogenic fungal species of plants and the fungi that had agricultural, military, and industrial importance. Among early taxonomists we also can list in chronological order, Fred J. Seaver (Rogerson, 1973), H.C. Greene (Backus and Evans, 1968), Dorothy I. Fennell (Hesseltine, 1979), and Travis E. Brooks (Keller, 1979). Most early mycologists were dedicated to the study of the moulds or filamentous fungi. Herman J. Phaff (1986), on the other hand, dedicated his career to the taxonomic study of the yeasts. In his autobiography, Phaff traced the development of his career according to the different developmental stages of microbiology that occurred during his life as a scientist.

II. The era of discovery: 1894 to 1919

Figure 6. Thomas Caspar Gilchrist (pp. 15–17) (Courtesy of Johns Hopkins Medical Archives)

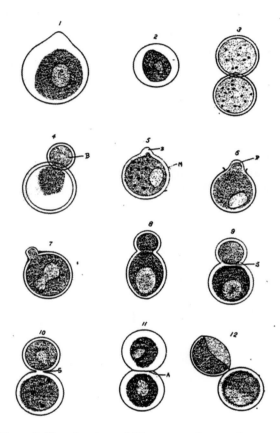

Figure 7. First drawings of *Blastomyces dermatitidis* (p. 15) (Gilchrist & Stokes, Circa 1896)

11

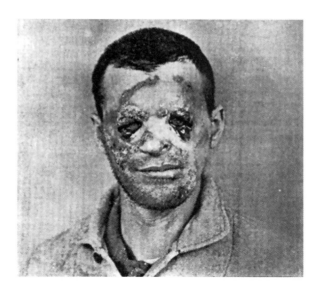

Figure 8. Second *Blastomycosis* patient (pp. 15–16) (Gilchrist & Stokes, Circa 1896)

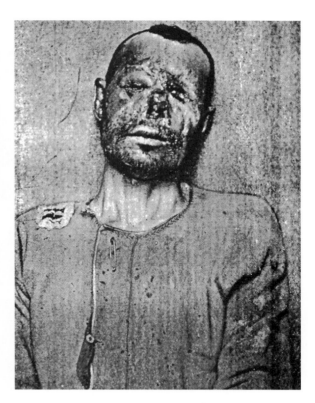

Figure 10. First *Coccidioidomycosis* patient in the United States (p. 16) (Rixford & Gilchrist, Circa 1896)

Figure 9. (opposite) Emmet Rixford (Circa 1896) (p. 16) (From: Derensinski, S. & Hector, R. 1996. *Coccidioidomycosis*. Washington, DC: NFID))

12

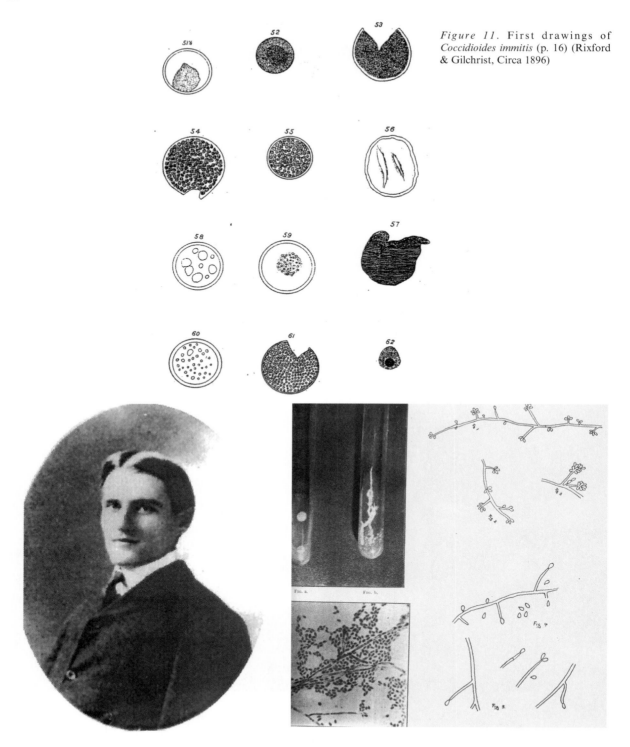

Figure 11. First drawings of *Coccidioides immitis* (p. 16) (Rixford & Gilchrist, Circa 1896)

Figure 12. Benjamin R. Schenck (1872–1920) (p. 17) (Courtesy of Johns Hopkins Medical Archives)

Figure 13. First drawings of *Sporothrix schenckii* and photo of first culture (p. 17) (Schenck, Circa 1898)

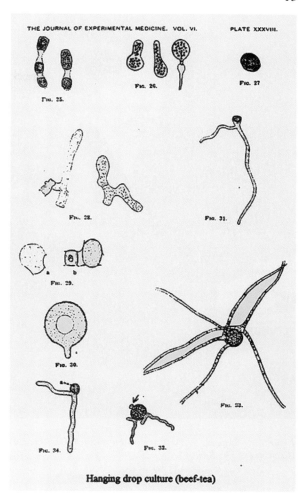

Figure 14. William Ophüls (1871–1932) (p. 18) (From: Derensinski, S. & Hector, R. 1996. *Coccidioidomycosis.* Washington, DC: NFID)

Figure 15. Dimorphism- *C. immitis* (p. 18) (Ophüls, Circa 1905)

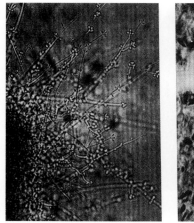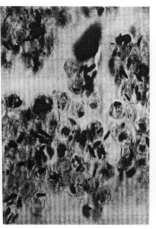

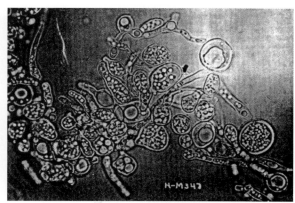

Figure 16. Dimorphism- *S. schenckii* (p. 19) (Ludvig C. F. Perkins, Circa 1900–1901)

Figure 17. Dimorphism- *B. dermatitidis* (p. 19) (Hamburger, Circa 1907)

Figure 18. Samuel T. Darling (Circa early 1900s) (p. 20)
(Courtesy of Dr. Ajello)

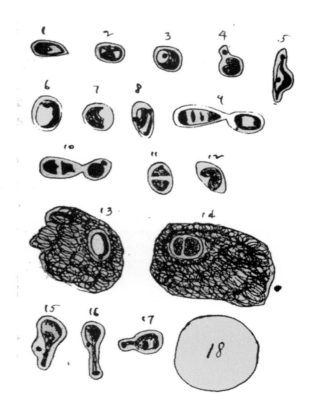

Figure 19. First drawings of *Histoplasma capsulatum* (p. 20)
(Darling, Circa 1906)

Figure 20. Henrique da Rocha-Lima (p. 20) (From: Sweany, H.
C. (Ed.). 1960. *Histoplasmosis*. Springfield, Il: Courtesy of
Charles C. Thomas)

Figure 21. William H. Welch (Circa late 1890s) (p. 21) (Courtesy
of Johns Hopkins Medical Archives)

Overview: 1894 to 1900

The development of medical mycology in the United States began during the late 1800s with the discovery of two new etiologic agents of disease: *Blastomyces dermatitidis* in 1894 and *Sporothrix schenckii* in 1899. Thomas Caspar Gilchrist and Benjamin R. Schenck conducted a thorough investigation of these mycoses at the newly established (1889) medical center of Johns Hopkins Hospital in Baltimore, Maryland. Research at Johns Hopkins was initiated in its new bacteriology laboratory with the support of federal grants. These major breakthroughs were soon followed by the discovery of the first two American cases caused by the dimorphic fungus *Coccidioides immitis* that were reported in 1896 from California. The prevalence of dermatophyte infections (as ringworm) in the Boston area and the first report of eumycotic mycetoma were published in 1898 and 1899, respectively. These early contributors were either dermatologists, pathologists or bacteriologists studying fungal infections, who followed the bacteriologic standards set by Louis Pasteur's research on rabies in France and Robert Koch's discoveries regarding tuberculosis in Germany. The publications describing fungal diseases originating from Europe and Latin America gave the early American investigators comparative reference points that became another contributory factor in the origin of the discipline. By the late 1800s, most American publications were illustrated with numerous drawings and photographs, and consisted of lengthy and detailed descriptions of lesions, tissue reactions, culture results, and experimental infections dealing with the etiologic agents.

The recognition of fungi as etiologic agents of disease is considered an important contribution to the development of the discipline by 30% of the contemporary medical mycologists, who responded to my questionnaire regarding the foundation of the discipline (Table 4, Appendix A). The thoughts of two respondents nicely summarize the discoveries of the late 1800s. Karl Clemons (1994) stated that "if fungi had not been proved to be disease carrying, we would not have a field of medical mycology". For Demosthenes Pappagianis (1994), "the recognition by the Europeans in the 1830s of the relationship between fungi and disease in humans represents the basis on which the further development of medical mycology depended" (K. Clemons, 1994; D. Pappagianis, personal communication, 1994).

1894 to 1900

Blastomycosis

The first American report of a fungal etiologic agent of human disease appeared in 1894, following Thomas Caspar Gilchrist's presentation of a blastomycosis case at the meeting of the American Dermatological Society held in Washington, D.C. in June of that year. Gilchrist (1896) later published an extensive and detailed account of this case, including numerous drawings and photographs of the tissue form (the yeast) of the etiologic agent that Gilchrist and W.R. Stokes named *Blastomyces dermatitidis* in 1898. This etiologic agent was first seen when Gilchrist (1894, 1896) was examining hand skin tissue for the presence of *Mycobacterium tuberculosis*. The patient lived in Philadelphia and was under Louis A. Duhring's care. The chief hematological features were numerous, well-defined, miliary abscesses of different sizes containing polymorphonuclear cells (PMN), nuclear debris, a few epithelial cells, scarce giant cells, and some budding yeasts. The walls of the abscesses consisted of epithelial cells; large collections of endothelial cells and PMNs were seen around the blood vessels.

B. dermatitidis yeast cells in tissue were described by Gilchrist (1896) as being deeply-stained, homogeneous, oval and roundish cells (6 to 16 μm) with a well-marked spherical nucleus (3.5 to 4 μm) that stained more deeply than the rest of the yeast cell. He described what he thought was a capsule as the "peripheral zone or the space between the protoplasmic contents and the membrane" (Gilchrist, 1896, p. 273). The developmental stages were studied from single cell to budding cell and the separation of the daughter cell from the parent cell. Each definition was accompanied either by drawings, photographs, or both. In 1894, *B. dermatitidis* was initially considered to be a protozoan by Gilchrist, but the presence of budding yeast cells suggested to him in 1896 that he was actually dealing with a yeast. However, he believed erroneously that this yeast was *Cryptococcus neoformans,* which had been reported previously in Europe by Otto Busse, A. Buschke and F. Sanfelice between 1894 and 1895 under the name *Saccharomyces neoformans* (Busse, 1894; Buschke, 1895; Sanfelice, 1894) (See Chapter I).

In 1896, a 31-year-old man with multiple skin lesions was referred to Halstead at the Johns Hopkins Hospital for evaluation. Gilchrist and Stokes (1896, 1898) then had the opportunity to study more closely a second case of blastomycosis. The infection had begun eleven years earlier as a small pimple located on the back of the patient's left ear, which subsequently became pustular. The disease process progressed very slowly (3 to 5 cm in 4 to 5 years), and gradually several lesions covered almost his entire face. The central portion of each lesion presented as an atrophic "cicatricial" condition and the whole area was enclosed by a distinctive and irregular border of demarcation. The planned treatment was to curette the active cutaneous lesions and then to apply silver nitrate; however, the patient left and later reported the spontaneous healing of these painless lesions without additional treatment. In 1896, Gilchrist provided only a drawing of the facial lesions, but later in 1898 included a photograph of the man's face showing the severity of the disease. He also included drawings of hyphae, yeast cells, and the tissue reaction. Sections from the cutaneous lesions showed the presence of budding yeasts (10–20 μm) and tubercles like those seen in the first case.

From the miliary abscesses, Gilchrist and Stokes (1896, 1898) obtained pure cultures within one week, which later developed profuse mycelium. The fungus grew on plain agar, potato agar, and beer-wort agar. The hanging drop culture was the technique used to observe the development of hyphae, which was described as many "long branching threads like mycelium, which are divided into shorter hyphae by intervening fine transverse lines" (Gilchrist and Stokes, 1898, p. 63). Conidia arose from the hyphae on short conidiophores in older cultures. Dogs, guinea pigs, a horse, and a sheep were successfully inoculated with hyphal cells and the tissue reaction in the experimental animals was identical to that seen in the human patients. The disease was named "blastomycetic dermatitidis". Gilchrist and Stokes (1898) recommended examining tuberculous-like lesions of the skin for "blasto-mycetes". The yeasts could be easily missed in tissues when stained with hematoxylin and eosin stains (H&E stains) (Gilchrist and Stokes, 1898) because of the unfamiliarity of most pathologists with the fungus in tissue and poor staining of tissue sections.

Coccidioidomycosis

In 1896, Emmet Rixford and Thomas Caspar Gilchrist reported two cases of what they thought were unquestionably genuine parasitic infections not to be ranked with doubtful "protozoa or blastomy-cete cases". The disease, coccidioidal granuloma, had been reported in 1892 by Alejandro Posada (Chapter I). The patients were two Portuguese men who had lived in California in the San Joaquin Valley for several years. One was admitted to a San Francisco hospital in July 1893 and the other to St. Mary's Hospital in August, 1894. The latter case was presented to the California Academy of Medicine on September 15, 1894. Both patients had similar cutaneous lesions involving the face, neck and extremities that were "horribly destructive". One patient lost both eyes, his upper lip, and half of one ear prior to death. The second patient died before his skin lesions became as extensive as those of the other patient and, unfortunately, no autopsy was performed in his case. The autopsy of the first patient revealed involvement of the lungs, adrenals, genitals, liver, spleen, and bone. The tissue reactions were described as having "close resemblance, amounting to histological identity, between many of the nodules and a genuine tuberculous process with caseation or coagulative necrosis" (Rixford and Gilchrist, 1896, p. 245).

Rixford and Gilchrist (1896) described the tissue form of the etiologic agent of coccidioidomycosis as "spherical, unicellular bodies, varying from 7 to 27 μm in diameter and consist in the encapsulated condition of a thick doubly countered capsule which encloses a finely granular protoplasm; as many as 100 small spores are set free by the bursting of the capsule" (p. 246). Lesions of experimentally infected animals showed caseation and large numbers of the fungal cells. The attempted cultures were assumed to be negative, because they were mistakenly discarded as being contaminated by a filamentous fungus. They gave the name, "protozoan or coccidioidal pseudo-tuberculosis" to the disease, and designated the organism from the first case as *Coccidioides immitis* and as *C. pyogenes* from the second case. They were not certain if the two fungi were the same, because one patient had a chronic process and the other an acute disease. Numerous illustrations and photographs were provided to support the descriptions of the lesions, tissue reaction, and the developmental stages of the pathogen in tissue.

Sporotrichosis

Benjamin Schenck (1898) reported that on November 30, 1896, a patient presented himself at the Surgical Clinic of the Johns Hopkins Hospital with an infection of the right arm. The primary infection involved the index finger, with the infection extending up the radial side of the arm, following the lymph system, and giving rise to several uncircumscribed, ulcerated indurations. Examination of skin tissue revealed "the characteristics of a chronic abscess, consisting of inflammatory and cicatricial tissues, a zone of leukocytes and newly-formed connective tissue, in which are several minute secondary abscesses" (Schenck, 1898, p. 288). In contrast to Gilchrist's (1898) fungus, the etiologic agent was reported to stain well and to be gram positive. Cultures of the "gelantinous, puriform material" and of small pieces of tissue gave rise to abundant growth of an "organism, not resembling bacteria". Growth appeared at 48 h (37°C) and mature colonies (72 h) were described as "opaque, white, and moist, with well-defined edges". Experimental infections were successful in a dog and mice.

Schenck (1898) used the hanging drop technique to study the development of the fungus in culture and then described how the "conidia germinate by sending out one or more unbranched and straight germ tubes from one end that give off spores; the spores are attached by terminal or lateral short pedicles or sterigmata" (p. 287). Schenck acknowledged Erwin F. Smith, United States Department of Agriculture, for the description. It was Smith who placed this fungus in the genera *Sporotrichum*. Smith believed that the genera *Sporotrichum* and *Trichosporium* were the same in contrast to Saccardo's opinion. Smith could not identify the fungus to a species level based on Saccardo's limited descriptions of more than 100 species (Schenck, 1898) of *Sporotrichum*.

Mycotic mycetoma

The first well documented case of a mycetoma in the United States was provided by James Wright in 1898. Based on a review of the literature, Wright concluded that mycetomas were caused by at least two different fungi. One caused an infection characterized by the presence of black granules and the other by pale or ochroid granules. Wright's (1898) patient was a 26-year-old Italian woman who was admitted to Massachusetts General Hospital on December 29, 1897 for evaluation of a possible six-month-old mycetoma. The foot (amputated) was swollen with "desquamation of the epithelium" and with a small sinus tract that "exuded on pressure a dirty, greyish fluid, containing black, hard, irregular granule-like grains of gunpowder" (Wright, 1898, p. 426). The granules were treated with KOH and crushed under a cover-glass to reveal "ovoid or rounded translucent bodies" cut in various planes and of varying sizes, typical septate hyphae, and atypical connective tissue. The fungus grew on various solid media in 4 to 5 days and appeared first as a "tuft of delicate, whitish filaments", which increased in number and length and did not bear conidia. On potato agar the colonies became dark brown and moist. The fungus was considered to be a hyphomycete.

Dermatophytosis

The last major medical mycology scientific paper published in the 1800s was by Charles White (1899), a dermatologist, who reported his personal observations of 279 cases of ringworm that he had treated in Boston between October 1895 to July 1898. Of the 279 cases, 139 (50%) were caused by *Microsporum audouinii* in children under 13 years of age, 127 by a mould that he called "megalospora", and 13 were not diagnosed. The plants as White (1899) called these etiologic agents were located on the scalp (120 cases), on the scalp and neck (3), in the beard (1), on the face (5), the trunk (3), and on the extremities (3) of 80 males and 59 females. The treatment for these infections consisted of thorough surgical measures for easily treated cases or epilation followed by application of either sulphur, napthol, carbolic acid, mercury, or curetting of the entire area and later skin-grafted for "obstinate disease". White (1899) concluded that American infections were milder than the European cases and required less severe treatment. However, microscopically, clinically, and culturally these ringworm infections resembled the ones reported in London and Paris, especially in the former city.

Overview: 1900 to 1919

Important contributions to the development of medical mycology in the U.S. were achieved during this period (1900 to 1919). The concept of dimorphism was established for three fungi between 1900 and 1907: *C. immitis*, *Sporothrix schenckii* and *B. dermatitidis*. Studies on the biology, epidemiology, and serology of *C. immitis* were begun. An understanding of the epidemiology and serology of the dimorphic fungi was considered important in the development of medical mycology (Table 4, Appendix A). *B. dermatitidis* and *C. immitis* were considered distinctive etiologic entities of invasive disease. It was noted that it was essential to consider these mycoses in the differential diagnosis of various diseases, including tuberculosis. Three additional fungi were recognized as etiologic agents of disease during this period, *Histoplasma capsulatum* var. *capsulatum*, *Phialophora verrucosa*, and *Absidia corymbifera* (as *Mucor corymbifer*). In addition, the first two American cases of cryptococcal meningitis were reported from New York. Kyung Joo Kwon-Chung (1994) stated that "while medical mycology in Europe was still centered around dermatophytosis and other cutaneous pathogens, medical mycology in the United States expanded the studies of the deep mycoses" (K.J. Kwon-Chung, March 14, 1994). Although the discipline was established, there was much confusion and organized training was not available. Medical mycologic research expanded and knowledge was disseminated through publications of scienific discoveries and presentations at medical conferences.

1900 to 1919

Coccidioidomycosis

The physicians, William Ophüls and Herbert Moffit (1900) isolated a mould at autopsy from the lungs of a farm laborer who had entered the City and County Hospital (San Francisco) on January 26, 1900 and died on February 6, 1900. The patient had a painful and inflamed joint, irregular fever ("up to 104°C"), cough with mucopurulent and occasionally blood-stained expectorations, and swelling over the left eye. The lungs showed signs of irregular nodular consolidation with necrotic centers (1 cm), several abscesses, and pneumonia. The disease had spread to the spleen, kidneys, liver, bones, and lymph nodes. The fungus was seen in all infected tissues and grew on agar as a "white to slightly yellowish mould" that was visible at 48 h. The infection was reproduced in

guinea pigs by inoculating the mould into their veins. Although the animals did not die, the parasitic form of the fungus and typical tubercle-like nodes were found in their lungs. These results suggested to them that "the protozoon-like bodies and the mould were different stages in the development of the same fungus" (Ophüls and Moffit, 1900, p. 1472). The concept of dimorphism thus was introduced for the first time for *C. immitis*. A parasitic form occurs in tissue and a saprophytic form in nature or when the fungus is grown at room temperature on various media.

In 1905, Ophüls investigated further the etiology of coccidioidomycosis, a disease that was then thought to be caused by the two species designated by Emmet Rixford and Caspar Gilchrist (1896) as *C. immitis* and *C. pyogenes*. Ophüls (1905) reviewed the six reported cases and conducted growth studies in experimentally infected animals and on artificial media. He reported that in infected rabbits the parasitic form (spherules) arose directly from the arthroconidia (saprophytic infectious cells) by a process of enlargement. He concluded that the infection was not caused by a protozoan, but by a pathogenic fungus; the primary infection may be either cutaneous or pulmonary; the lesions ("infectious granulomata") contain chronic abscesses and nodes resembling tuberculosis; the fungus is pathogenic for animals (dogs, rabbits, and guinea pigs); and produced lesions in animals that were similar to those encountered in humans. He suggested the name *Oidium coccidioides* for the fungus and coccidioidal granuloma for the disease.

By 1915 Ernest Dickson concluded that 35 of the 40 known patients with coccidioidomycosis had been residents of California, three had visited the state, and 27 had spent some time in the San Joaquin Valley (16 had been evaluated at the Leland Stanford Junior University, School of Medicine, formerly Cooper Medical School). Most of the patients (37 of 40 patients) were adult male laborers. The review of these 40 cases prompted Dickson (1915) to conclude that coccidioidal granuloma was a clinical entity; the disease was almost identical to tuberculosis; coccidioidal granuloma and blastomycosis were etiologically distinctive diseases; coccidioidal granuloma may be treated by radical removal of infected organs; and some patients undergo spontaneous recovery. Therefore, there was the need to include these diseases in the differential diagnosis of various infections and to perform microbiologic identification of their etiological agents. During that same year, at the Unversity of California Hospital in San Francisco, Jean Cooke (1915) described the first immunological studies on coccidioidomycosis. She demonstrated the presence of precipitin antibodies (titer of 16) in the serum of a patient using an antigenic extract (as the "precipitinogen") of dried cultures of *C. immitis*. The serological reactions were specific, since the precipitin antibodies did not cross-react with *B. dermatitidis* antigen. Although she failed to demonstrate both complement-fixing (CF) and agglutinin antibodies, she suggested the potential use of the CF test as a diagnostic tool. W.B. Bowman (1919) was the first to describe the X-ray findings of five patients afflicted with this disease, while he was at the Los Angeles County Hospital. Photographs of bone and lung lesions were provided in his article.

Sporotrichosis

While at Rush Medical College, Shenandoah, Iowa, Ludvig Hektoen and C.F. Perkins (1900–1901) diagnosed a third case of *S. schenckii* infection and, most importantly, compared the characteristics of their isolate with those recovered by Benjamin Schenck (1898) and Brayton (Hektoen and Perkins, 1900-1901). Hektoen and Perkins (1900–1901) demonstrated that the three infections had been caused by the same fungal species, and concluded that the yeast form was the only form found in the tissues of humans and experimentally infected animals. This paper contains photographs and drawings that clearly depict the morphological characteristics of the saphrophytic and parasitic forms of this species. By 1913, David J. Davis had confirmed Hektoen and Perkins' (1900–1901) earlier suggestion concerning the dimorphism of *S. schenckii* when he studied Hektoen's, Gougerot's (Paris), and other clinical isolates originating from patients with sporotrichosis. Maurice Langeron, Gougerot, and other investigators had studied numerous cases caused by *S. schenckii* at the Pasteur Institute.

Blastomycosis

James Walker and Frank Montgomery (1902), a surgeon and a dermatologist, respectively at Rush Medical College, Shenandoah, Iowa, were the first to report that *B. dermatitidis* could cause invasive disease in humans. Their case was reported at a surgical meeting as a patient having only a skin infection. They originally believed that their patient had died of miliary tuberculosis which was the patient's secondary infection. Following a careful search for *M. tuberculosis* in several hundred skin sections, only a few bacilli were found in contrast to the large numbers of yeast cells that had been apparently missed by the pathologists. The lungs also were full of typical yeast cells (as blastomycetes), which clearly demonstrated that this was a systemic fungal infection. The patient had been curetted, each ulcer was cauterized with Pacquelin cautery followed by a daily application of wet boric-acid dressings. Nine days later, the clean granulating areas had been covered with Thiersch's grafts furnished by a friend of the patient. Nine days after the skin grafting, the patient was worse, new lesions were seen, and death occurred 33 days after surgery. It was believed that curettage hastened the demise of the patient, however, it was the indicated treatment when large doses of potassium iodide proved to be ineffective.

The cultural characteristics of *B. dermatitidis* were first described by Walter Hamburger (1907) when he reported that *B. dermatitidis* grew well and rapidly on routine laboratory media (e.g., glucose agar, glycerin-agar, ox blood serum agar, and potato agar) and even faster on slightly acid glucose-agar. However, his most significant contribution was his description of the dimorphism of this pathogen, when he concluded that room temperature favored production of mycelium and aerial hyphae and that at 37°C there was inhibition of hyphal growth and production of yeasts instead. The yeast-like colonies reverted to hyphal growth at ambient temperature, and the hyphal growth to yeast- like colonies at 37°C. Hamburger (1907) recommended the incubation of cultures at 37°C and ambient temperature, a practice currently followed in many clinical laboratories. The prevalence of systemic blastomycosis was confirmed by Frank Montgomery and Oliver Ormsby (1908) during the ensuing year. They "collected" and reviewed 22 cases (some unreported) from the Chicago area (13 cases), Iowa (2 cases), and from other areas of the United States. They also reviewed cases reported from Germany and France. The common and most pronounced feature of these cases was the formation of multiple abscesses in various organs of the body.

Histoplasmosis

Samuel T. Darling (1906), while he was working as a pathologist in Panama, found numerous small, oval to round bodies in autopsy smears of the lungs, spleen, and bone marrow of a Panamanian patient. This patient had been suspected of having miliary tuberculosis of the lungs. Most of these bodies were intracellular (10 to 100 fungal cells per host cell) within alveolar epithelial cells in the pseudotubercular areas, while others appeared to be free in the spleen and rib marrow. Polymorphonuclear leukocytes were rare, a few mononuclear cells were seen, and *M. tuberculosis* bacilli were absent. The "bodies were surrounded by a clear refractive non-staining rim, in thickness about 1/6 the diameter of the parasite" (Darling, 1906, p. 1283). Darling (1906) believed erroneously that this clear area was a capsule when he named this microorganism *Histoplasma capsulatum*. He thought it was a protozoan and not a fungus, since it resembled *Leishmania donovani*. Nevertheless, the discovery of another fungus as an etiologic agent of invasive disease had been accomplished. However, it was the Brazilian, Henrique da Rocha-Lima (1912–1913), who concluded that Darling's organism was in reality a fungus.

Chromoblastomycosis

The first known case of a *Phialophora verrucosa* infection was reported simultaneously in the United States by physicians in the Departments of Pathology (Medlar, 1915) and Dermatology (Lane, 1915) of Boston City Hospital. Two important diagnostic observations were made. First, laboratory diagnosis is needed for certain types of chronic skin infections and, secondly, all "fungi that appear as yeast cells in tissue are not necessarily of the same variety of fungus" (Medlar, 1915, p. 508). C.G. Lane's (1915) patient was a nineteen year old Italian immigrant, who for a year had had a painless lesion on the right buttock. The lesion had started as a purulent pimple that gradually increased to 2.5 cm in size. A painless lump developed near the lesion six months later. A diagnosis of tuberculosis verrucosa was considered and the lump was excised. The initial laboratory diagnosis was blastomycosis based on the "bodies seen in the smears." However, because the lesion appeared to differ from those known to be caused by either *B. dermatitidis* or *S. schenckii*, a biopsy was performed by Lane (1915) and then sent to E.M. Medlar (1915) for further study.

The biopsy revealed the same types of cellular reactions as those seen in blastomycosis, but the parasitic cells were phaeoid and occurred solitarily or in small groups. The fungal cells were described by Medlar (1915) as either "brownish-black sclerotic cells (8 to 15 µm) that show septations after maturation (size of mature cells, 20 to 25 µm) or as budding yeasts" (Medlar, 1915, p. 513). The microscopic observations of the isolates of this phaeoid (as black) mould, which was initially discarded as a contaminant, showed mycelium composed of septate branched hyphae supporting conidial formation on lateral branches by sequential budding, and "the formation of a shallow cup, where conidia are deposited, as the wall of the germinating cell is extended outwards" (Medlar, 1915, p. 515). Medlar (1915) gave his fungus to the famed mycologist Roland Thaxter at Harvard University, who concluded that it should be described as a new taxon. Medlar followed Thaxter's recommendation, used Saccardo's classification system and proposed the new name *Phialophora verrucosa*. Excellent photographs of the phialides and phialoconidia of this species, as well as of the parasitic forms of the fungus, appeared in both articles.

Zygomycosis

A brief and inconclusive report of an *Absidia corymbifera* (as *Mucor corymbifer*) infection of the vocal cords was reported in 1918 by H.C. Ernst from Harvard University. This appears to be the first report of zygomycosis caused by a fungus belonging to this genus in the United States. Experimental disease was reproduced in guinea pigs and it was found that the optimal temperature for growth was ambient temperature. The significance of this report is that *A. corymbifera* belongs to the group of fungi called Zygomycetes (as Phycomycetes), and that some of these species are important and prevalent etiologic agents of zygomycosis. It is noteworthy that such an unusual case of zygomycosis was the first to be described.

Training and education contributions: 1894 to 1919

During "the era of discovery", the early contributors to the development of medical mycology as a discipline were not trained as medical mycologists. The first detailed descriptions of infections caused by *B. dermatitidis* (Gilchrist, 1894), *C. immitis* (Rixford and Gilchrist, 1896), and *S. schenckii* (Schenck, 1898) were made by physicians who were students or mentees of William H. Welch at Johns Hopkins Medical School in Baltimore, Maryland. This medical center is often considered the leader in medical educational thought (Turner, 1976) because of the creative vision and imagination of its founders. Through private funding the University was able to support full-time clinical teachers as well as research programs in these early years. An emphasis was placed on the training of individuals as original investigators in order to continually increase medical knowledge and develop a corps of future scientists. Other important educational contributions during this era were the establishment of a bacteriology laboratory at Columbia University by Welch, before he moved to Johns Hopkins, and the introduction of the study of bacterial diseases into the formal curriculum of a botany course at the University of Illinois by Thomas J. Burrill (Espinel-Ingroff, 1994). By the late 1800s, Johns Hopkins already had established its reputation as one of the top medical research and training institutions in the United States and later became known as the teacher of teachers (Turner, 1976). During those early years, medical students at Johns Hopkins were required to conduct research projects and write a dissertation.

Welch encouraged interdisciplinary research and training and he was an important influence on students at that institution. The isolation of *S. schenckii* was conducted by Schenck (1890) while he was a medical student under the tutelage of Welch, but Schenck became a gynecologist after his graduation in 1898 (Schwarz and Baum, 1964). The appointment of William Ophüls as Professor in the Cooper Medical College in San Francisco in 1898 was made at Welch's suggestion. Ophüls became Professor of Pathology at Stanford University in 1909, where the study of coccidioidomycosis had begun in 1894. He had obtained his medical degree in Germany and was Johannes Orth's assistant for two years; Orth had been a student of Rudolph Virchow. Ophüls continued his research well into the 1900s. Samuel T. Darling, who discovered histoplasmosis capsulati in the Panama Canal Zone in 1905, was trained as a pathologist at the College of Physicians and Surgeons in Baltimore, whereas, Thomas Caspar Gilchrist was trained as a dermatol-

ogist in England, where he had been born. Gilchrist, who conducted his laboratory investigations in Welch's laboratory, examined the patient's specimens and made the first diagnosis of blastomycosis. Gilchrist's accurate drawings were reproduced for many years in several textbooks. He was probably the first individual to teach undergraduate medical students the morphologic methods of diagnosis in dermatology while he was a professor of dermatology at both Johns Hopkins and the University of Maryland (Schwarz and Baum, 1964).

In the second half of the nineteen century, American medicine saw the emergence of laboratory science as part of the practice of medicine and the acceptance of non-physician scientists into the field of medicine. The origin of bacteriologic research and the decision to educate medical students in bacteriology during the "era of discovery" was vital to the development of medical mycology and related sciences. However, during the ensuing years (1900 to 1919), medical mycology continued being of concern primarily to physicians; a phenomenon that also had been observed in related medical fields. As American physicians became aware of the importance of Pasteur's and Koch's research contributions in Europe (Walsh and Poupard, 1989), bacteriology became a necessary part of medical training. As physicians realized the importance of this training, bacteriology became a component of the medical curriculum and the study of fungi was considered by many to be part of the "bacteriology exercise".

Although accounts of the histories of the infectious disease specialty (Kass, 1987), internal medicine (Beeson, 1986), the Infectious Diseases Society of America (Kass and Hayes, 1988; Kunin, 1988), and tropical medicine (Warren, 1990) had been published, none of these historians attributed any merit to or mentioned the contributions made by medical mycologists to the study of infectious diseases during this era. Recent historical studies regarding key early figures in microbiology (Chernin, 1987; Dolman, 1982; Rapp, 1989; Woodward, 1989) only gave credit to the individuals who had made advances to understanding the importance of tuberculosis, diphtheria, tetanus, and cholera in public health. Prior to these discoveries, physicians considered the presence of bacteria accidental in tissue. These contributions marked the late 1800s as the "golden era of microbiology".

However, as early physicians diagnosed new cases of systemic fungal diseases, they became interested in the management of the mycoses and investigation of these new etiologic agents of infection. In their search to better understand the mycoses that mimicked tuberculosis, these early researchers rapidly expanded knowledge of the pathogenic fungi and fungal diseases. Schwarz and Baum concluded in 1964: "Great admiration, obviously, must be expressed for the intuition and power of observation of these men [and women] who had rather poor optical instruments, little background on which to build and no specialists to consult" (p. 80).

III. The formative years: 1920 to 1949

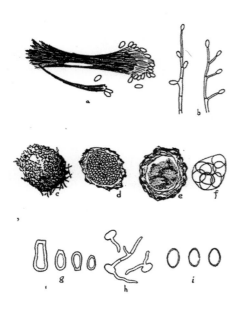

Figure 22. First drawings of *Pseudallesceria boydii* (*Scedosporium apiospermum*) (pp. 33–34) (Boyd & Crutchfield, Circa 1921)

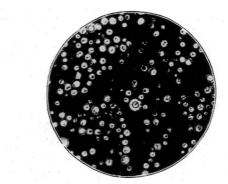

Fig. 7.—Cells from two months' old culture showing mucoid capsules. India ink method.

Figure 23. First cryptococcal meningitis in the United States (p. 34) (Freeman & Weidman, Circa 1923)

Figure 24. J. Gardner Hopkins (p. 35) (Circa early 1920s) (Courtesy of Dr. Silva-Hutner)

Figure 25. First Medical Mycology Laboratory (1926) and Rhoda William Benham (p. 35) (Courtesy of Dr. Silva-Hutner)

24

Figure 27. M.E. Hopper at the Vanderbilt Clinic (late 1920s) (p. 37) (Courtesy of Dr. Silva-Hutner)

Figure 26. Rhoda Williams Benham (1894–1957) (pp. 35–37) (Courtesy of Dr. Silva-Hutner)

Figures 28 and 29. Back and front of Rhoda Benham medallion (p. 37). (From: MMSA Archives).

Figure 30. Chester Wilson Emmons (1900–1985) (pp. 37, 45) (Courtesy of Dr. Kwon-Chung)

Figure 31. Myrnie Gifford (Circa 1930s) (pp. 37–38) (From: Derensinski, S. & Hector, R. 1996. *Coccidioidomycosis*. Washington, DC: NFID)

	Fresno	Kern	Kings	Madera	Merced	San Joaquin	Stanislaus	Tulare	Totals
Number of doctors reporting cases	19	10	7	6	4	12	7	10	75
Number of cases reported	69	117	24	31	11	29	18	55	354
Males.....................	27	48	6	9	4	15	5	25	139
Females.................	42	69	18	22	7	14	13	30	215
Fever....................	68	117	17	30	11	27	15	52	337
Cough and sputum.......	19	48	6	22	1	4	9	15	124
Recovery without complication...............	65	115	20	31	11	15	17	51	325
Progressed to typical coccidioidal granuloma	..	1	..	..	..	..	..	..	1

Figure 32. Valley Fever Cases from 1936 to 1937 (pp. 37–38) (Dickson & Gifford, 1938)

Figure 33. Katharine Dodd (p. 38). (From: Sweany, H. C. (Ed.). 1960. *Histoplasmosis*. Springfield, Il: Courtesy of Charles C. Thomas)

Figure 34. William A. DeMonbreun (p. 38). (From: Sweany, H. C. (Ed.). 1960. *Histoplasmosis*. Springfield, Il: Courtesy of Charles C. Thomas)

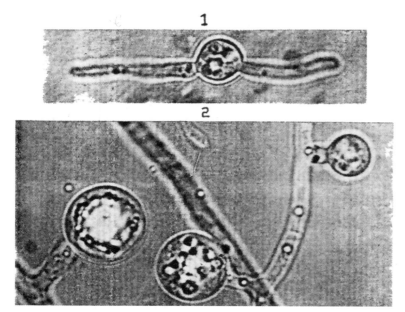

Figure 35. Dimorphism-*H. capsulatum* (p. 38) (DeMonbreun, Circa 1934. Am. J. of Trop. Med. XIV, p. 23)

Figure 36. Norman Francis Conant (1908–1984) (pp. 39, 46) (Courtesy of Dr. Campbell)

Figure 37. Charles E. Smith (1904–1967) (pp. 39–41) (Courtesy of the School of Public Health, University of California at Berkeley

Figure 38. Amos Christie (pp. 39, 41). (From: Sweany, H. C. (Ed.). 1960. *Histoplasmosis*. Springfield, Il: Courtesy of Charles C. Thomas)

Figure 39. Carroll E. Palmer (pp. 39, 41). (From: Sweany, H. C. (Ed.). 1960. *Histoplasmosis*. Springfield, Il: Courtesy of Charles C. Thomas)

Figure 40. Charlotte Campbell (first row, 1st) (p. 42). Virginia & Charles D. Jeffries (second row, 4th & 5th) (p. 60) (Courtesy of Dr. Jeffries)

Figure 41. William H. Weston (1890–1978) (p. 43–44) (Courtesy of Dr. Silva-Hutner)

Figure 42. William H. Weston (top row, 4th), Mentor of Silva-Hutner (second row, 1st) and other students (1948) (pp. 43–44) (Courtesy of Dr. Silva-Hutner)

Figure 43. Rhoda Benham and her mentee, Arturo Carrión, at Columbia in 1949 (p. 44) (Courtesy of Dr. Silva-Hutner)

Figure 44. Orda A. Plunkett (p. 44–45) (Courtesy of Dr. J. Shadomy)

Figure 45. From left: Michael Furcolow (second), Chester W. Emmons) & Norman Francis Conant (Courtesy of Dr. Goodman)

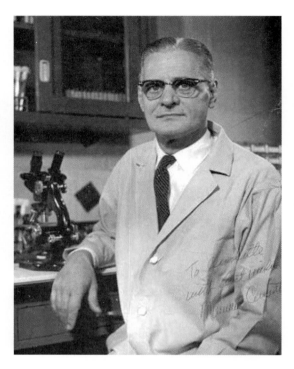

Figure 47. Norman Francis Conant (pp. 39, 46) (Courtesy of Dr. D. MacKenzie)

Figure 46. First Medical Mycology Course by Norman Francis Conant (third row, 5th), 1947. Duke University (pp. 39, 46) (Courtesy of the Duke University Medical Center Archives)

31

Figure 48. Howard W. Larsh (1914–1993) (pp. 40, 47) (Courtesy of Dr. Goodman)

Figure 49. Howard Larsh's students in the 1970s: Larsh (first row, 4th), Nancy Hall (3rd), Glenn Roberts (5th) and Norman Goodman (7th) (pp. 40, 47) (Courtesy of Dr. Goodman)

Figure 50. Libero Ajello (pp. 40, 47–48)

Overview: 1920 to 1929

The direction of medical mycology was affected by the occurrence of a number of scientific events during the 1920s. The first medical mycology laboratory was established in 1926 at Columbia University. Serological and cell mediated studies were conducted with crude antigens of *C. immitis* and *Cryptococcus neoformans*; diagnostic laboratory tests began to be developed; *Pseudallescheria boydii* (as *Allescheria boydii)* was identified as an etiologic agent of disease (mycetoma); revision of reported cases and isolates were conducted as a means of classifying different isolates within a species or genus; better descriptions of histoplasmosis capsulati and its etiologic agent were published; and sporotrichosis was declared an occupational disease. Serologic and other diagnostic tests were among the major contributions singled out by 9% and 10.1%, respectively of the questionnaire respondents (Table 4, Appendix A). Development of skin tests and the work of Raimond Jacques Sabouraud were considered the most important contributions by 7% of the 86 individuals who answered the question regarding these issues (Table 3, Appendix A). For Jose Vazquez (questionnaire), "Sabouraud's book, *Les Teignes*, on the dermatophytes is one of the first descriptions of fungal diseases and their etiologic agents".

1920 to 1929

Laboratory diagnostic methods

Bertha Fineman (1921) stated that bacteriological textbooks in the beginning of the twentieth century had "meager" descriptions of the etiologic agent of candidiasis, known as thrush during that time period (Department of Bacteriology and Immunology, University of Minnesota, Minneapolis). The fungal taxonomists of the era had given *C. albicans* at least 10 different names, since Charles P. Robin (1853) had named this organism *Oidium albicans* following its discovery by B. Langenbeck (1839) in a case of typhoid fever and in 1890, W. Zopf introduced the name, *Monilia candida*. Because carbohydrate fermentations and agglutination tests were useful in the 1920s to classify bacteria, Fineman (1921) adopted this system to study what she called the "thrush parasite". She grew cultures on "dextrose-tartaric agar" and several other differential media (e.g., milk,

gelatin, and carrots), tested 17 carbohydrates, and adopted the medium developed in Europe by Linossier and Roux for chamydoconidium formation. Three important conclusions were made by Fineman (1921) concerning *C. albicans* (as *Oidium albicans*): carbohydrates are useful for yeast identification; isolates of this species produce chlamydoconidia but no ascospores; and this species tends to be filamentous in liquid media at low oxygen and surface tensions, and unicellular on solid media.

Two important contributions were made concerning the selection of a suitable medium for studying dermatophytes at the University of Alabama, Birmingham (Hodges, 1928) and the University of Pennsylvania, Philadelphia (Weidman and Spring, 1928). The peptone manufactured by Difco (Detroit, Michigan) was found to be unacceptable as a substitute for the imported French peptone. The morphological characteristics, that are distinctive for each dermatophyte species, were not evident when using the American peptone. This created a problem, because their identification relied heavily on Sabouraud's descriptions and illustrations of the dermatophytes grown on agar containing the French peptone. After numerous experiments had been performed by Fred Weidman and Dorothy Spring (1928), using a variety of medium formulations supplemented with peptones and carbohydrates, a suitable medium was selected by Robert Hodges (1928). The final formulation of this medium contained an American peptone (1%, Fairchild), dextrose (4%), agar (1.8%), and had a final pH of 5.0. This formulation was named Sabouraud medium (liquid and agar) and to this day it is the basic routine culture medium used to grow fungi in most clinical laboratories.

Mycotic mycetomas

Mark Boyd and Earl Crutchfield (1921) described the first case of mycetoma caused by *Pseudallescheria boydii* (as *Allescheria boydii*) at the University of Texas Medical Branch in Galveston, and reviewed 30 cases of mycetomas caused by other etiologic agents. The mycological studies of *P. boydii* were performed by C.L. Shear (1922). The patient's foot had been infected following trauma from a thorn. Three months later, the ankle became "soft and the lesion finally ruptured, discharging bloody pus" (Boyd and Crutchfield, 1921, p. 239). The lesions healed temporarily but continued to develop for 12 years and contained yellow-white granules that were composed of a dense homogeneous network of mycelial "threads". Shear's subcultures of the dark fungus grown on cornmeal and glycerin agars developed three forms of "fructifications": conidia, thin-walled globose to subglobose ascospores in cleistothecia, and synnemata. The colonies were white at first, later becoming grayish-black with the conidia being borne on loose-branched hyphae. Pier Antonio Saccardo had classified a similar fungus, identified in 1893 by M.J. Constantin, under the generic name *Allescheria*. Shear (1922) named this fungus *Allescheria boydii* in honor of Boyd. The treatment of the era for mycetomas was solely surgical; both local and systemic therapy were not effective. Surgical treatment involved either removal of small areas of tissue or amputation of the extremity, the latter being considered the "best course" (Boyd and Crutchfield, 1921). *P. boydii* was the first fungal pathogen known to reproduce sexually.

Cryptococcosis

In 1923, Walter Freeman and Fred Weidman at the University of Pennsylvania reported a case of "cystic infiltration" of the brain caused by a "blastomycete" that they identified as *Cryptococcus neoformans* (as *Torula histolytica*). The name *T. histolytica* had been given by J.L. Stoddard and E.C. Cutler in 1916 (New York, NY, Rockefeller Institute of Medical Research) to a yeast they had found in infected tissues of two American patients. A similar infection had been reported in three European patients as a "true *Torula* infection" (cryptococcal meningitis). However, Freeman and Weidman (1923) dismissed Stoddard and Cutler's two cases as unacceptable *T. histolytica* infections, recognizing only the three European cases and the yeast recovered by Langdon Frothingham (1902) from the lungs of a horse. Because Stoddard and Cutler had not cultured their patient's yeast, Freeman and Weidman (1923) conducted an extensive study of their own case. This may be considered the first well-documented case report of this disease in the United States. The patient, aged 39, was admitted to the hospital with symptoms of severe paroxysmal headaches of one month duration, vomiting, poor vision, rigidity of the neck, and difficulty in writing and talking. He had had a prior diagnosis of Hodgkin's disease in 1916. The patient worsened during the following two

months and then died. Yeast cells were easily seen in the cerebro-spinal fluid (CSF), and their capsules were demonstrated by making an "emulsion of yeasts" in India ink, a rapid diagnostic test that is still in use. The yeast was grown on blood agar and Loeffer's blood serum at 37°C at room temperature and was described as being spherical with double contoured walls and variable in size. Stained tissue sections showed edema, fibrous hyperplasia, and cellular infiltrates with plasma cells and lymphocytes, but without PMNs. The yeasts were found both intra- and extracellularly; only a few of the cells were budding. The "outstanding abnormality was the presence of numerous cysts that varied markedly in size" (Freeman and Weidman, 1923, p. 594). Although physiologic studies were performed, no conclusive results were reported.

By 1926, the pathologist B.Z. Rappaport and the bacteriologist Bertha Kaplan reviewed 13 reported cases of cryptococcosis (2 German and 11 American patients) and concluded that cryptococcal meningitis was acquired through the respiratory system by inhalation of *C. neoformans*, which disseminated from the lungs to other organs. These authors found two significant clinical features in their patients, the presence of skin lesions that were different from those caused by *B. dermatitidis* and a greater degree of dissemination than previously thought. They also noted that although iodides, tartar emetic solution, and intraspinal injections of immunized rabbit serum and colloidal silver were used as therapy, the infection was always fatal. Rappaport and Kaplan (1926) were the first to serologically test the serum and CSF from a patient with cryptococcosis as well as sera from immunized rabbits and guinea pigs by complement fixation (CF) and agglutination tests.

Histoplasmosis capsulati

In 1926, William Riley and Cecil Watson (University of Minnesota) reported the fourth case of histoplasmosis capsulati and the first in the United States. Samuel Darling had described the first three cases between 1905 and 1906 in Panama at the Ancon Hospital that represented three cases out of 33 000 admissions (Darling, 1906; Riley and Watson, 1926). Riley and Watson (1926) had the opportunity to study their patient antemortem: a German woman who had lived in the United States for 42 years, having had a cough for more than 28 years and who died four weeks after admission. She had developed an acute respiratory condition with severe dyspnea and cyanosis and had similar clinical symptoms and postmortem findings as found in Darling's patients (1906). Using the Giemsa stain, Riley and Watson (1926) recognized that the microorganism that they saw did not have the characteristic features of the *Leishmania* species; however, as Darling (1906) had concluded before them, they thought the microorganism was encapsulated. Therefore, they erroneously concluded that histoplasmosis was caused by two related species of *Cryptococcus* (as *Grubyella*), one causing systemic infections and the other superficial infections (Riley and Watson, 1926). Histoplasmosis was recognized as a fungal disease, but because cultures had not been obtained, the real identity of the etiologic agent of this disease remained unknown.

Skin test development

Edwin Hirsch and Harriet Benson (1927) were the first to successfully obtain skin test reactions in patients suffering from coccidioidomycosis. They used heat-stable (80°C), soluble filtrate antigens from 14-day-old peptone broth cultures of *C. immitis*. The skin reactions resembled those seen with tuberculin, and the immediate skin reaction was similar to the "wheal" produced by specific pollen extracts used in patients with hay-fever. Prior attempts at skin testing by Jean V. Cooke (1915) and others on animals or on patients had given conflicting results because the hypersensitivity reactions were not specific for *C. immitis*. The report by Hirsch and Benson (1927) may be considered the forerunner of the more intensive studies of skin testing performed during the 1940s.

Sporotrichosis

The dermatologist Harry Foerster (1926) reported 18 cases of sporotrichosis during a five year period. Since most of his patients worked at plant nurseries, and the disease was acquired in at least 10 of the 18 patients following trauma from barberry thorns, he concluded that sporotrichosis was an occupational hazard among individuals engaged in farming and horticulture. From that time on, American courts of law began to award compensation for the cost of medical care for this disease as an employment-connected disability (Ainsworth, 1986).

First research and training center

The first medical mycology laboratory in the United States (Columbia-Presbyterian Medical Center, New York) was established in 1926 by Professor J. Gardner Hopkins, a dermatologist and Zinsser-trained microbiologist. By 1950 over 4000 clinical specimens were being processed annually (37% positive) at this laboratory (Benham, 1950b). While bacteriological laboratories were abundant around the country, there were no laboratories devoted to the etiologic diagnosis of fungal infections. Until then, most medical mycologic studies had been conducted by physicians and bacteriologists in bacteriology laboratories. Hopkins began carrying out his own culture work from skin lesions while Bernard O. Dodge, a professor of Botany at Columbia University in New York, NY, helped him with fungal identifications. Because they became aware of the prevalence of fungal infections, they recognized the need to have a full-time mycologist. Robert Harper, a Professor of Botany at Columbia University and Chester Emmons' major professor, recommended Rhoda Benham as Hopkins' full-time assistant. Benham stated in 1950, that although "it was the first time I had ever heard that fungi caused human disease, I landed the job" (p. 1). In 1926, Benham was conducting research for her doctoral degree under the supervision of the classical mycologist H. Richards ("Training Trees", Appendix B) (Silva and Hazen, 1957). The Rockefeller foundation awarded a five-year grant ($50 000) for the support of research in medical mycology to Columbia University on March 11, 1929 (Rockefeller Foundation, 1929). This represented the first grant awarded for research and training in medical mycology.

Overview: 1930 to 1939

By the beginning of the 1930s, there was no clear distinction among the three important systemic fungal diseases: blastomycosis, coccidioidomycosis, and cryptococcosis. Also, an inadequate classification system was being used to identify *C. albicans* from other yeasts. In addition, the dermatophytes were classified according to Sabouraud's system, which was primarily based on the clinical presentation of the infection. Rhoda Benham and Chester Emmons at the Medical Mycology Laboratory, Columbia University provided a more logical classification system for the identification of these fungi. By the end of the 1930s, the incidence of fungal diseases involving the central nervous system and the prevalence of coccidioidomycosis as a mild disease in the San Joaquin Valley had been reported. The first report of a *Fonsecaea pedrosoi* infection in the United States, the isolation of *C. immitis* from the soil, and the establishment of *H. capsulatum* var. *capsulatum* as a dimorphic fungus appeared in the literature between 1932 and 1936.

In 1935, Norman F. Conant was invited to join the staff at Duke University (formerly Trinity College until 1930) where the third case of chromoblastomycosis caused by *F. pedrosoi* was waiting for him, as he stated in 1969 (Conant, 1969). Another important contribution to the development of medical mycology was the recognition by the Federal government of the need to establish a Mycology Section at the National Institutes of Health (NIH) in Bethesda, Maryland. This was accomplished with the appointment of Chester W. Emmons as the senior mycologist. By 1939, the first three medical mycology centers that had been established were located in the eastern region of the United States.

1930 to 1939

Taxonomy and classification

The 1930s began with the publication of an excerpt of Benham's doctoral dissertation in 1931. This classical paper shed much light on the very confusing subject of yeast identification. The confusion at that time was due to the lack of a generally accepted classification system, as well as recognized criteria for the identification of yeast species that were pathogenic to humans. Most investigators had referred to these fungi as either simply blastomycetes, or by a number of different generic names. The "rudimentary" morphology of these fungal pathogens and their varied appearance under different cultural conditions complicated matters.

In order to end the confusion surrounding the species of *Candida*, Benham (1931) systematically investigated their morphologic characteristics, carbohydrate fermentation reactions by using Durham's peptone and tubes, pathogenicity, and agglutination reactions (absorbed serum). She selected 30 yeast strains recovered from thrush (19 isolates), sputa, feces, and plants. For her morphological

studies, Benham (1931) used corn meal agar and Dalmau's methodology of placing a cover slip "directly on the growth". Data from this procedure permitted her to conclude that the majority of the fermenting, yeast-like fungi, isolated from thrush lesions, were identical and belonged to the well defined species *C. albicans* (as *Monilia albicans*). The other species of yeasts that she studied had "definite morphological differences", which Benham depicted clearly in her excellent drawings. The 19 strains identified by her as *C. albicans* were pathogenic for rabbits and some of these strains had "similar antigenic content as determined by reciprocal absorption of agglutinins" (Benham, 1931, p. 215). Agglutination reactions, using absorbed sera and performed in conjunction with several other tests, seemed to be her most useful identification criteria. Even today, the Dalmau technique is a good identification tool for *Candida spp*. Six years after Benham's (1931) first paper, a survey of the "mycologic flora of 1000 adults' normal mouths and throats" was undertaken as a means of determining the role of pathogenic yeasts in thrush infections. It was found that *C. albicans* was present in 7% of "normal" individuals and that 22.5% of their sera had agglutinins against this yeast (Todd, 1937).

Benham's paper concerning the classification of *C. albicans* was followed by two important contributions in 1934 and 1935. In 1934, she studied three different mycotic infections, blastomycosis, coccidioidomycosis, and cryptococcosis, which questionably had been called yeast infections. Benham (1934) definitely stated that each infection was a distinct disease that could be diagnosed by the characteristic morphology of their etiologic agents in tissue and culture. In 1935, Benham studied more than 40 isolates of the genus *Cryptococcus*, which included the original cultures from Otto Busse, A. Buschke, and F. Curtis, as well as strains recovered from the skin and intestinal tract of either normal individuals or patients suffering digestive diseases (Benham, 1935). She concluded that the species of the genus *Cryptococcus* may be divided into four groups based on cultural and serological characteristics as well as animal pathogenicity. However, her most significant conclusion was that all of the pathogenic isolates belonged to only one species, *C. neoformans* (as *Cryptococcus hominis*) (Benham, 1935).

Benham's studies were not confined to these genera. She became interested in the nutrition of fungi when she discovered that *Malassezia furfur* (as *Pityrosporum ovale*) required oleic acid to grow in vitro. Benham (1939) later studied the vitamin nutritional requirements of *M. furfur*, thus introducing the study of fungal nutrition, a field in which several other investigators became interested at that time. Benham's publications on the nutrition of dermatophytes, many of them in collaboration with her students Lucille K. Georg and Margarita Silva-Hutner, enlarged the knowledge of medical mycology in this country (M. Silva-Hutner, personal communication, November 20, 1993). Silva-Hutner was her eventual successor. In 1969, in recognition of her achievements, the Medical Mycological Society of the Americas (MMSA, founded in 1966) established the Rhoda Benham Award to be presented at the Society's annual meetings to individuals who had contributed significantly to the development of medical mycology. The choice of Benham as the award's name bearer was based on her pioneering role in the development of medical mycology throughout the Americas (Ajello, 1969).

In 1929, Chester W. Emmons was offered a position as Associate in Mycology by Professor J. Gardner Hopkins at Columbia University. Emmons had little training with human pathogens, but he took the opportunity to expand his knowledge of fungal pathogens while he managed the Department of Dermatology's culture collection. At that time, the collection consisted mainly of dermatophytes, a few yeasts, and isolates of *S. schenckii*. He collected other dermatophyte isolates from patients' skin, hair, and nails at the Vanderbilt Clinic, and eventually used these isolates for his classical taxonomic study of the dermatophytes. The result of these two "extracurricular" experiences was exhibited in 1932 at the American Medical Association's annual meeting on *Pleomorphism and Variations in the Dermatophytes* (Campbell, 1983). More importantly, his scholarly review of the literature led him (1934) to realize that there was a "well recognized need in medical mycology for a more logical and usable classification of the dermatophytes according to the botanical rules of nomenclature" (p. 337). Emmons (1934) followed the development of "conidiophores and conidia" (as spores) by using Henrici's modified slide culture in which agar was run under a cover slip fastened to a slide; he substituted sterile shavings of cows' horns for the agar. Based on morphological characteristics, Emmons (1934) proposed retention of the generic names *Trichophyton*, *Microsporum*,

and *Epidermophyton* and declared that the genera *Achorion* and *Endodermophyton* were unnecessary. Emmons (1934) believed that "such a natural mycological classification follows in a general way Sabouraud's clinical classification and that the inconsistencies of the latter classification can be avoided" (p. 362). His proposed classification system remains in use today. In February 15, 1936, the arrival of Emmons at the NIH began his successful career there. The establishment of the Clinical Mycology Service at the NIH not only provided financial support but, most importantly, increased the visibility of the discipline and directed scientific progress.

Coccidioidomycosis

A brief report appeared in the literature in 1932 that described the first isolation of *C. immitis* from "earth" collected around the sleeping quarters at a ranch near Delano, Kern County, California. The "fungus spores" were set free from the soil using a column of brine (Stewart and Meyer, 1932). Between 1937 and 1938 three papers were published by individuals at Stanford University, Stanford, California (Dickson, 1937a,b; Dickson and Gifford 1938). Significant information was provided from these publications concerning *C. immitis*, because the authors had the fortunate opportunity of studying five patients infected with *C. immitis* exhibiting the primary infection known as "valley fever" (of unknown etiology then). "Valley fever" was characterized by colds or bronchopneumonia, high temperature, painful erythema nodosum or skin lesions. One of the five patients was a medical student, who had acquired the infection in the laboratory when he inadvertently opened a "petri dish culture" of *C. immitis* that was several months old and dried. Later he stated that he had observed "a light-brownish cloud arising from the dish, a cloud of chlamydospores" (Dickson, 1937a, p. 151). The other patients had acquired the infection on their jobs in California's San Joaquin Valley, where they had been exposed to a great deal of dust. Although some of these patients at first were diagnosed as having pulmonary tuberculosis by X-ray, the coccidioidomycosis diagnosis was established by the demonstration of the typical spherules in sputa, with subsequent culture confirmation (Dickson, 1937b).

It became evident then that the first manifestations of the fungal disease caused by *C. immitis* were either an acute "cold", bronchopneumonia, erythema nodosum, or all three. It was also confirmed that inhalation of the arthroconidia (as chlamydospores) was followed by a primary infection and in some cases secondary progressive disease (Dickson, 1937b; Dickson and Gifford, 1938). Ernest Dickson and Myrnie Gifford (1938) suggested that descriptions of coccidioidomycosis should include the different clinical manifestations of this disease. The fact that "valley fever" was actually one of the manifestations of coccidioidomycosis, demonstrated that this fungal disease was not rare.

Meningeal infections

The prevalence of meningeal infections caused by several fungal pathogens was reported in New York, NY and Washington, DC by Walter Freeman (1933) and Lawrence W. Smith and Machteld Sano (1933). Freeman (1933) studied tissue sections from 13 cases of *C. neoformans* meningitis and one case each of CNS infections caused by *Aspergillus* spp., *B. dermatitidis*, *C. immitis*, *H. capsulatum*, and *S. schenckii*. He concluded that these were chronic meningeal infections without invasion of other organs and that the finding of the fungus in the spinal fluid (CSF) established the diagnosis. At the same time, Smith and Sano (1933) reported what they considered to be the first case of *C. albicans* meningitis. The patient had a rapidly spreading oral infection with invasion of blood vessels, development of fungemia followed by localization of the infection in the meninges. This patient was a 22-month-old infant who had died nine days after admission at the Willard Parker Hospital. The yeast was recovered from the CSF and the lesions in his mouth. Benham from Columbia University's College of Physicians and Surgeons helped with the mycological studies and she identified the yeast as *C. albicans* (as *M. albicans*).

Histoplasmosis capsulati

Three significant publications regarding the *capsulatum* variety of *H. capsulatum* appeared in the literature in 1934 from the Vanderbilt University School of Medicine, Nashville, Tennessee (Dodd and Tompkins; DeMonbreun) and Georgetown University, Washington, DC (Hansmann and Schenken). Katharine Dodd and Edna Tompkins (1934) described the third case of histoplasmosis capsulati

in the United States (the sixth in the world). This was the first report describing this fungal disease in an infant, a six-month-old white male, as well as the first case diagnosed antemortem. The clinical and pathological findings, and the "morphology" of the fungus within the large mononuclear cells of the blood were similar to the cases that had been reported by Samuel Darling (1906) and William Riley and Cecil Watson (1926). The studies of the fungus were performed by William A. DeMonbreun (1934), who cultured material from the child on various media, was surprised when a mould was recovered, because only yeast cells were present in tissue. However, his further studies demonstrated that the yeast form "grew well in serum and blood agar cultures incubated between 34 to 37°C and the mould form in duplicate cultures incubated at ambient temperature (25 to 28°C)" (p. 101). The dimorphic nature of *H. capsulatum* was thus established. DeMonbreun (1934) also provided excellent descriptions and photographs of the diagnostic macroconidia of *H. capsulatum*. Macroconidia were described as spherical or pear shaped cells, 10 to 15 µm in diameter, that were attached to the end of the short hyphal branches. The thick, highly refractive cell walls of the macroconidia had 11 to 6 µm rounded projections, a crenated appearance, and resembled "Teutonic war clubs" (DeMonbreun, 1934). He referred to these "aerial spores as tuberculated spores".

G.H. Hansmann and John R. Schencken (1934) reported a case of a chronic infection caused by "a yeast-like organism in tissue" which they erroneously thought belonged to the genus *Sepedonium*. The culture had been studied by Chester Emmons while he was at Columbia and he identified it as a *Sepedonium* sp. based on the presence of the "large, spiculated, thick-walled, tuberculated chlamydospores" characteristic of that genus. Although they saw similar *H. capsulatum* "organisms" to those described by Darling (1906), Hansmann and Schenken (1934) did not believe that this was a case of histoplasmosis because there was no involvement of the spleen in their patient and their isolate was a mould. Emmons stated during his lecture at the annual MMSA meeting in 1968, "since the hyphal form of *H. capsulatum* was unknown to me at the time, I made the mycological reasonable identification of *Sepedonium*, it was fortunate that DeMonbreun was able to establish the dimorphism of *H. capsulatum*" (Emmons, 1968).

In Williamson County, Tennessee, R.S. Gass et al. (1938) observed by X-rays the remarkable prevalence of tuberculosis lesions in 24.7% of 1291 individuals, aged 6 to 19 years, who had negative tuberculin reactions. Their conclusion was that the level of tuberculosis was higher than the one actually detected by the tuberculin test. Later, Amos Christie and J.C. Peterson (1945) and Carroll Palmer (1945) found that these cases were caused by *H. capsulatum* var. *capsulatum*, an endemic fungus in that area of the country, and not by *M. tuberculosis*.

Chromoblastomycosis

In 1936, the first case of a *Fonsecaea pedrosoi* (as *Hormodendrum pedrosoi*) infection in the United States was reported by Donald S. Martin, Roger D. Baker, and Norman Francis Conant of Duke University in Durham, North Carolina. This case of chromoblastomycosis (as dermatitis verrucosa, a name coined earlier by Brazilian investigators) had began four years earlier on the left hand and fingers of the patient as a series of pruritic vesicles and pustules, which progressed to elevated, scaling, verrucous nodules with itching and burning and with such a tremendous amount of swelling that the patient's arm was supported by a sling. Their examination of the slide cultures demonstrated three types of conidiogenous cell formation: "a *Hormodendrum* type (known now as *Clasdosporium*-like) in which 3 to 6 μm in length and 1.5 to 3 μm in width conidia (as spores) were borne at the apex of the conidiophore as branching chains; an acrotheca-like (known now as a sympodial form) conidial head in which the conidia appeared as tooth-like processes on the apical portion of the terminal cell of some of the conidiophores; and a *Phialophora* type in which the conidia were formed in flask-shaped conidiophores" (Martin, Baker, and Conant, 1936, p. 600). The authors gave credit to Arturo Carrión and Chester Emmons from the University of Puerto Rico, who had provided them with their unpublished manuscript describing the presence of the *Phialophora* type of conidiogenous cells in cultures of *Fonsecaea* (as *Hormodendrum*). This document facilitated the morphological studies performed at Duke University. Also, it was reported from Duke University (Swartz and Conant, 1936) that the treatment of skin scrapings with 5% KOH, followed by washing with water and staining with a lactophenol-cotton blue preparation facilitated the detection of fungi in clinical materials, thus avoiding the various confusing artifacts found when using only KOH. Lactophenol-cotton blue is one of the stains used in clinical laboratories for staining culture preparations.

Fungal endotoxins

In 1939, Arthur Henrici at the University of Minnesota, Department of Bacteriology, reported for the first time that *Aspergillus fumigatus* produced two endotoxins: a hemolytic toxin and a potent pyrogenic toxin. The importance of Henrici's discovery was recognized later during the well known "turkey" epidemic in England in the early 1960s. The turkeys, swine, poultry, and cattle affected with the severe systemic disease had consumed foodstuffs that contained aflatoxin, a metabolite produced by *Aspergillus flavus*.

Overview: 1940 to 1949

As a consequence of World War II, many new aspects of coccidioidomycosis and histoplasmosis capsulati were discovered. Investigations by the physician, Charles E. Smith, led to studies on the epidemiology of coccidioidomycosis, and later histoplasmosis capsulati. These findings led Emmons to become involved in fungal ecological studies that resulted in his isolation of *H. capsulatum* var. *capsulatum* from soil in Virginia. As Norman Goodman stated in 1994, "the development of skin testing gave us the ability to differentiate between infection and disease and conduct epidemiologic studies, and opened the doors for the development of better data on the natural history of the systemic mycoses". Understanding the epidemiology of fungal infections removed much of the mystery from these opportunistic diseases as leading individuals came together and investigated the full range of cause and effect.

Reports of cases of *Candida* spp. endocarditis and meningitis, *P. boydii* meningitis, and rhinocerebral zygomycosis began to appear in the literature between 1940 and 1948. The identification of fungi was advanced by the development of G. Gomori's stain for tissue sections, by Roger D. Baker's observations that different types of tissue reactions could be observed in the mycoses, and by the development of carbohydrate assimilation tests for yeast identification.

Charles E. Smith, was recruited in 1934 by Ernest Dickson, a professor of Public Health in the Stanford School of Medicine to work on *C. immitis*. From Stanford, Smith moved to the University of California School of Public Health at Berkeley, where he founded another center for medical mycologic research and training ("Training Trees", Appendix F). Between 1940 and 1945, the Armed Forces found it necessary to maintain a mycology laboratory at the Walter Reed Army Medical Center in Washington. In this laboratory, Charlotte Campbell, Samuel Saslaw, and G. Hill performed thousands of serologic tests for the diagnosis of blastomycosis, coccidioidomycosis, and histoplasmosis capsulati. A. Wadsworth, Director of the New York State Laboratory in Albany, appointed Elizabeth Hazen in 1944 as the first diagnostic and research mycologist for this organization (Bacon, 1976). The increase of fungal diseases during World War II had stimulated the creation of this position (Gordon, 1993). Because Hazen was working in Benham's laboratory in New York at the time, samples for fungal examination were sent from Albany, where the State laboratory was located (Gordon, 1993). Hazen had joined the State Laboratory as a bacteriologist in 1931, and received her training in medical mycology at Benham's laboratory ("Training Trees", Appendix B). Following the epidemiologic studies with *H. capsulatum* var. *capsulatum*, the federal government established the Kansas City Field Station as part of the Atlanta, Georgia's Center for Disease Control (CDC) to study non-tuberculosis calcified pulmonary disease in the Mississippi-Ohio River Valley area. Under the direction of the physician, Michael Furcolow, between 1945 and 1964 this center became the source of "most knowledge about histoplasmosis capsulati and, as a result, other groups became involved with the study of fungus infections" (Conant, 1969; Furcolow, undated).

In 1946, the CDC hired Libero Ajello and sent him first to Duke University to start the CDC's Medical Mycology Unit there with Norman Conant. But two months after Ajello's arrival at Duke, the United States Public Health Service decided that the CDC Medical Mycology Unit should be in Atlanta. Ajello developed an excellent team of mycologists at the CDC that worked on all aspects of medical mycology: laboratory diagnosis, training, and research (W. Kaplan, personal communication, January 6, 1994). The Medical Mycology Unit that Ajello established at the CDC later became the

Division of Mycotic Diseases. This mycologic diagnostic and applied research laboratory served the nation's needs, and soon Ajello's group began to present short training courses for medical technologists and physicians, and established a serology program (Chezzi, 1992; L. Ajello, personal communication, January 4, 1994).

In 1945, Howard Larsh began his graduate program at the University of Oklahoma at Norman and in 1947, the physician, John Utz, initiated his clinical studies and training program at the NIH. By the end of 1949, medical mycologic research and training were also being conducted at two other centers: Michigan State University by Everett Beneke and the University of Cincinnati by Jan Schwarz and Gerry Baum (Espinel-Ingroff, 1994, 1996).

According to the Annals of The New York Academy of Medicine (1950), the first medical mycology conference in the United States was held by the Biology Section of the New York Academy of Medicine from October 31 to November 1, 1947 in New York City. The monograph entitled, *Medical Mycology*, which contained the series of papers presented at this scientific meeting, was published in 1950 in The Annals of the New York Academy of Sciences.

1940 to 1949

Epidemiology and ecology

Charles E. Smith, while at Stanford University, reported in 1940 the first cooperative field and laboratory study of San Joaquin fever ("valley fever") in Kern and Tulare counties, California, where this infection was endemic. During a 17 month period (December 7, 1937 to May 12, 1939), 432 patients, who recovered without sequelae, were enrolled in the study. In addition to the clinical picture and the presence of erythema nodosum or erythema multiforme, a positive coccidioidin (filtrate of liquid cultures developed at Stanford in 1939) skin test also served as a criterion for enrolling patients in this study. The sensitivity to coccidioidin became evident two to seventeen days after the onset of symptoms. Many facts were learned and the methodology involved in the preparation of coccidiodin was published in 1948 by Smith and coinvestigators.

The seasonal incidence of coccidioidomycosis corresponded to the climate, with the peak in the dusty fall and an ebb in the wet winter; benign "valley fever" was more common among white females and severe granuloma among "dark-skinned" males; eventually, most individuals in the region experienced the infection – an estimated 8000 to 10 000 infections in the two counties, but only 5% of them developed erythema nodosum (Smith, 1940). This study made a significant contribution to the development of medical mycology, because it expanded and confirmed the results obtained by Dickson and Gifford between 1937 and 1938 concerning the significant prevalence of benign coccidioidomycosis in the western regions of the United States.

During World War II, medical mycology became important to the military, because thousands of non-immune troops were sent to receive training in the arid areas of California and Arizona, where *C. immitis* was highly prevalent and many trainees adquired coccidiodomycosis. Charles E. Smith became a civilian consultant to the Secretary of War to study and to assist with the control of coccidioidomycosis in the Armed Forces. Smith, Rodney Beard, H.G. Rosenberger, and E.G. Whiting's investigations for the Armed Forces were published in 1946, which included medical care, skin testing, serological tests, and laboratory identification. Detailed epidemiological investigations were carried out at four army field camps in the San Joaquin Valley from July 1941 until February 1946. The results of these studies emphasized the importance of dust control by "grassing, paving roads and airfield runways, and ultimately the use of highly refined oil on athletic areas" (Smith et al., 1946, p. 838). Smith also demonstrated that coccidioidin reactors had acquired immunity to the disease.

During the summer of 1941, *C. immitis* was isolated from 5 of 150 soil samples collected in the desert and from 25 of 105 rodents trapped near the village of San Carlos, Arizona (Emmons, 1942). The demonstration that a high percentage of Indian school children in this area were sensitive to coccidioidin and had demonstrable pulmonary nodes (Aronson, Saylor, and Parr, 1942) prompted Emmons' ecological studies. These two studies confirmed previous findings made during the 1930s that failure to react with tuberculin in patients having calcified nodes (as nodules) may be due to a fungus and that *C. immitis* was recovered from the soil in endemic areas.

Also in 1941, the first intradermal injections of *H. capsulatum* var. *capsulatum* soluble filtrate antigens (histoplasmin) were given to a patient and experimentally infected animals by Paul Van Pernis, Miriam Benson, and Paul Holinger. The antigen produced specific, immediate and delayed, cutaneous reactions which led to Amos Christie and J.C. Peterson's (1945) and Carroll Palmer's (1945) epidemiologic studies. Palmer, an officer of the U.S. Public Health Service, had been studying the geographical variation in the prevalence of pulmonary calcification and tuberculin non-reactivity. Christie, who had gained experience in California with the benign form of coccodioicomycosis, informed Palmer that such non-reactors were histoplasmin-positive.

Although Emmons was not a co-author of Palmer's (1945) publication, Emmons prepared the histoplasmin for Palmer by growing two strains of the *capsulatum* variety of *H. capsulatum* in the medium that was used by Smith (1940) to grow *C. immitis* for producing coccidioidin. Palmer (1945) tested histoplasmin and tuberculin in approximately 3000 student nurses living in Detroit, Michigan, Kansas City (Missouri and Kansas), Minneapolis, Minnesota and Philadelphia, Pennsylvania. Based on their skin test reactions and chest X-rays, Palmer stated in 1945: "mild, probably subclinical infection with *H. capsulatum* is widely prevalent in certain states, relatively infrequent in others and the incidence of *H. capsulatum* infection corresponds with the presence of pulmonary calcification in most tuberculin negative individuals" (Palmer, 1945, p. 519). Palmer confirmed Christie and Peterson's (1945) results, who had examined 181 histoplasmin skin test reactions and X-ray films in children at Vanderbilt University (Nashville, Tennessee). Although at the time, neither of these authors arrived at a definite conclusion, their studies provided evidence of the high prevalence of benign histoplasmosis in certain areas of the United States. These discoveries also led to the ecological studies by Emmons (1949), who was able to isolate *H. capsulatum* var. *capsulatum* from 2 of 387 soil samples collected between December 18, 1946 and December 18, 1948 from farms in Loudoun County, Virginia. Emmons previously had found that dogs and other animals trapped in the area had histoplasmosis capsulati. Two fatal histoplasmosis cases in human siblings had been diagnosed in Loudoun County.

Systemic candidiasis

Harry Joachim and Silik Polayes reported the first case of candidal endocarditis in 1940 from New York City (Cumberland Hospital). The patient was a 48-year-old white male, who had been addicted to morphine and heroin for twenty years. He had given himself intravenous injections without aseptic or antiseptic precautions for eighteen months. Joachim and Polayes (1940) stated that "the aortic vegetations had a layer of fibrin and exudate covering massive colonies of *Monilia*" (p. 207). Although this case was attributed to a species of *Candida* (as *Monilia*), the identity of the yeast was not given. Another case of candidal endocarditis and meningitis was reported in 1946 following massive treatments (four weeks each) with penicillin for two episodes of bacterial endocarditis (Geiger, Wenner, Axilrod, and Durlacher, 1946). The brain sections showed an intense inflammatory reaction in the meninges with many mononuclear cells and a few yeast cells. The yeast was identified as *C. albicans* (as *M. albicans*) by Emmons at the NIH. These were the first reports of severe opportunistic fungal diseases as a sequel to broad-spectrum antibiotic therapies, as well as the association of *C. albicans* with intravenous drug use.

Zygomycosis

Strikingly similar cases of rhinocerebral zygomycosis (as mucormycosis of the meninges and brain) were diagnosed in three patients with severe diabetes mellitus at the Johns Hopkins School of Medicine by John Gregory, Alfred Golden, and Webb Haymaker (1943). Upon admission the patients were unable to talk and developed increasing drowsiness that progressed to a coma. In the next few hours, their faces became edematous, particularly about the eyes, until the whole face was involved, with the patients then dying during the first to the fifth hospital day. Cultures were not obtained. Tissue sections from these cases were sent to Emmons at the NIH, who stated, "the large diameter of the hyphae, the manner of branching, and the coenocytic structure are characteristics of the *Mucoracea*, probably a *Mucor* sp." (Gregory et al., 1943, p. 411). Rhinocerebral zygomycosis became associated with acidotic diabetes following this well documented report in 1943. The first well documented case of rhinocerebral zygomycosis had been described in 1885 from Ger-

many by A. Paltauf. H. Bauer, Libero Ajello, E. Adams, and D.U. Hernandez (1955) were the first to isolate and describe the prime etiologic agent of rhinocerebral zygomycosis, *Rhizopus arrhizus* (as *R. oryzae).*

Eumycotic mycetomas

In 1944, Chester Emmons demonstrated that the fungus Shear (1922) had described as *Pseudallescheria boydii* (as *Allescheria boydii)* and the fungus described from numerous cases of mycetoma outside the United States, as *Monosporium apiospermum,* were in reality the same fungus. *Pseudallescheria boydii* "was the ascocarpic stage of *M. apiospermum*" (p. 192). The pathogenicity of *P. boydii* was found to be inherent in both the sexual and asexual forms by David Lupan and John Cazin (1973) while they were working at the University of Iowa. Rhoda Benham and Lucille Georg (1948) reported the first case of meningitis caused by *P. boydii* while J. Creitz and H.W. Harris described the presence of this fungal pathogen in a cavitary lung lesion in 1955.

Emmons (1945) described the first case of mycetoma of the hand caused by the phaeoid fungus *Exophiala jeanselmei* (as *Phialophora jeanselmei*). This fungus had been originally described by the French mycologist, Maurice Langeron, as *Torula jeanselmei.* Because Emmons (1945) erroneously thought that it resembled other members of the genus *Phialophora*, he transferred it to *Phialophora* as a new species.

Laboratory diagnostic methods

In 1947, Charlotte Campbell reported that Francis' glucose-cystine blood agar, a bacteriologic medium, "luxuriantly supported growth of the yeast phase" of *S. schenckii* (Campbell, 1947, p. 263), and that a modified formulation of this medium also supported the growth of *H. capsulatum* var. *capsulatum*'s yeast form. This medium was then used in clinical microbiology laboratories for conversion of this fungus from its mould form to its corresponding yeast form. This test permitted definitive identification of the etiologic agent of sporotrichosis.

During the 1940s, several investigators concentrated on determining the nutritional requirements for the growth and "spore" production of the dermatophytes. William Robbins and Roberta Ma (1945) discovered that *Trichophyton mentagrophytes* required a combination of either amino acids or vitamins as a source of nitrogen. Later, Elizabeth Hazen (1947) from the New York State Health Department demonstrated that yeast extract, when added to honey agar, resulted in an increase in the production of the macroconidia by *Microsporum audouinii*. Lucille Georg (1949), while at Rhoda Benham's laboratory, studied several species of *Trichophyton* and their nutritional "deficiencies". These preliminary studies led to Georg's other significant discoveries in the next decade. She also provided the first description of the macroconidia of *Trichophyton violaceum*.

Several scientific advances made between 1946 and 1948 facilitated the diagnosis of fungal infections in the laboratory. The first description of a stain for the microscopic observation of fungal cells in tissue was published by G. Gomori (1946), a pathologist from the University of Chicago. His histochemical test was based on the principle that fungal elements were stained black by a reduction of a silver solution on the cell wall of the fungus. In 1955, this stain was modified by R.C. Grocott and later by Milton Huppert, D.J. Oliver, and S.H. Sun (1978) to further optimize the detection of fungi in tissue. Using other stains, Roger D. Baker (1947) made extensive observations of the tissue reactions of a series of fungal infections and attempted to catalog them based on the degree of suppuration, macrophages, giant cells, caseous necrosis, and fibrosis. He observed all these tissue reactions in the invasive mycoses, but did not find caseous necrosis in chromoblastomycosis nor inflammatory responses in mycotic superficial infections. His conclusion was that "chronic suppuration with fibrosis was the most general tissue change in deep fungal infections. Neutrophils were usually the primary reacting cell but either macrophages or giant cells were also components of the host's primary tissue response" (Baker, 1947, p. 466). The observation of asteroid formation, an important tissue reaction to *S. schenckii*, was reported for the first time by Morris Moore and Lauren Ackerman (1946).

In 1948, the usefulness of assimilation tests in the classification of yeasts was demonstrated by Lynferd Wickerham and Kermit Burton (1948). Wickerham had spent several years evaluating fermentation and nitrogen assimilation tests to improve the techniques used in the classification of yeasts. Wickerham and Burton (1948) employed a better basal medium than others had done, and utilized more carbon compounds than evaluated by other investigators. The technique used was Beijerinck's modified procedure in which small amounts of the carbon compounds were placed on the inoculated agar. The test was based on the development of growth where the assimilable compounds were placed. These simple assimilation tests were modified and became an invaluable tool in clinical microbiology laboratories for many years, leading to the development of commercial kits based on this methodology during the 1970s. The tests avoided the complications of reversed fermentation reactions or transitory nitrate assimilation.

Antifungal agents

Reports on the antifungal activity of three different compounds: benzimidazole (Woolley, 1944), propamidine (Elson, 1945), and cycloheximide (as actidione) (Leach, Ford, and Whiffen, 1947) appeared in the literature after 1944. Benzimidazole was the first reported imidazole to inhibit the growth of the saprophytic yeast *Saccharomyces cerevisiae*. Low concentrations of propamidine were found to be fungistatic against *A. fumigatus*, *B. dermatitidis*, and *S. schenckii*. Cycloheximide (as actidione) was the compound added to a mycologic medium, mycobiotic agar or mycosel (Georg, Ajello, and Gordon, 1951). Although propamidine was used only briefly as a therapeutic agent in humans, this drug was a forerunner of the intense search for active and safe antifungal agents during the ensuing years.

Training and education contributions: 1920 to 1949

Harvard University

One of the important early contributors to medical mycology during "the formative years" was William H. Weston, who taught at Harvard University from 1921 to 1956. Known as "Cap", he was a classical mycologist whose love for the study of fungi stimulated a new generation of scientists, a group of pioneers who would lead "an era of unprecedented development in experimental mycology in the United States" (Sequeira, 1993, p. 43). He was trained by Roland Thaxter, the renowned mycologist who provided descriptions and superb illustrations of many different groups of fungi. He taught the young

Weston to heed the smallest details, a precept he would pass on to his students.

But what attracted so many bright young students to "Cap's" classes was his teaching ability. His enjoyment of his work and infectious humor provided what Sequeira (1993) calls, "an atmosphere where they (students) were allowed absolute freedom to pursue their interests and to follow their best instincts as to the way to approach and solve scientific questions" (p. 47). Although Weston did not overly stress medically important fungi in his courses, an impressive line of medical mycologists emerged from his learning environment that included Norman Conant, Margarita Silva-Hutner, Ralph Emerson, Frederick Wolf, Charles M. Wilson, Samuel B. Salvin, and Arden Howell ("Training Trees", Appendix B). These early leaders would, in turn, train many of the well-known medical mycologists, who have made important contributions to the discipline up to our current time (C.M. Wilson, 1979).

Columbia University

Recognition for the first formal medical mycology training program in the United States begins with J. Gardner Hopkins at Columbia University. In 1929, Hopkins utilized the Rockefeller Foundation grant to expand the Medical Mycology Laboratory into a teaching and research center (Rockefeller Foundation, 1929; Silva-Hutner, 1975). Thus, Hopkins and Benham had begun a 30-year partnership that would produce numerous significant advances in medical mycology and train many of the first generation of medical mycologists in the United States ("Training Trees", Appendix B). The laboratory soon became well known, attracting students from across the world: the first Ph.D. graduate was Lucille Georg. In the 1950s, Arturo Carrión came for training to Benham's laboratory. Although the elimination and redirection of federal funds terminated the formal training programs in the 1980s, the laboratory remained a strong research and diagnostic resource under the guidance of Irene Weitzman until her retirement in the 1990s. Traditionally, research centers and laboratories have provided and continue to offer training opportunities in the biomedical sciences. Wilson (1972) stated that the training of microbiologists shifted from centers of research in Europe, which had advanced the training of American scientists until the 1930s, to American research

centers after World War II. Following the second war, there was unprecedented support of research by the federal government through NIH grants in the health sciences (Watson, 1969).

Benham also is credited with starting the first accredited course in medical mycology in 1935. This three credit, 12 week course in the Medical School was taken by dermatology fellows and microbiology graduate students. Classes were given twice a week, including a one-hour lecture and two hours in the laboratory. A dedicated teacher, Benham was well organized and prepared for the course that covered all the medically important fungi known at that time. She was creative in the use of learning aids such as slide cultures, "lantern slides", histology sections, drawings, animal models of ringworm and cryptococcal infections, and a card reference system organized in her laboratory manual. The lectures were given by both Benham and Hopkins, the latter who was also a skilled lecturer and teacher (M. Silva-Hutner, personal communication, November 20, 1993). Benham was an avid photographer, and frequently used this medium in her professional and personal life. She helped during the laboratory sessions, involved her technologists as teaching assistants, and required the students to do their own drawings of fungal structures.

Benham and Hopkins worked together closely as a team until 1948 when Benham had a heart attack, a setback that kept her from work for nearly a year. She was able to resume her work for several more years, but in 1955, her health failed again and she was unable to return to her laboratory. She continued to write, completing several manuscripts until her death in 1957.

University of California at Los Angeles

The first undergraduate medical mycology course is believed to have been started in 1937 by Orda A. Plunkett at the University of California, Los Angeles (Carlyn Halde, personal communication, December 9, 1993). Plunkett was a botanist, who was very knowledgeable in fungal biology. His course devoted one semester to basic general mycology and one semester to medical mycology. In addition, he offered seminar courses on both subjects. The general mycology section covered the taxonomy of the ascomycetes and other fungi. The medical mycology section was given three days a week for six to eight hours. Learning aids for Plunkett's course included his extensive collection of tissue and culture slides, drawings, and cultures. A number of Plunkett's students belonged to the Microbiology Department, but conducted their thesis work in his laboratory for medical mycology (e.g., H. Jean Shadomy, Carlyn Halde, Frank Swatek, etc.; "Training Trees", Appendix B). Plunkett gave much of his time as a mentor to his students, often at the expense of promotions and his own career. He did not publish research results until he had obtained data from a number of studies. Another strong training component of Plunkett's courses was that he allowed his students to help in teaching the laboratory sections, thereby enhancing their knowledge and presentation skills (J. Shadomy, January 17, 1994).

Two other well-known contributors to medical mycology, who were colleagues of Plunkett, were John F. Kessel of the University of Southern California and J. Walter Wilson, a physician. Kessel spent a limited amount of time devoted to medical mycology, but is best known as an educator who gave guidance to other well known medical mycologists that included Wilson and E.E. Evans. Wilson was a practicing dermatologist in Los Angeles, who frequently was invited to give medical mycology lectures to medical groups (Howard, 1985).

The National Institutes of Health (NIH)

A major contributor to medical mycology and general mycology was Chester W. Emmons, whose work influenced the training of numerous medical mycologists. Emmons had earned his M.S. in botany under George W. Martin, the noted mycologist and taxonomist at the University of Iowa, Iowa City in 1927 (Kwon-Chung and Campbell, 1986; "Training Trees", Appendix B). Emmons moved to Columbia University where he received a Roberts Fellowship to pursue his Ph.D. with Robert Harper, a well known cytologist. Once again coincidence came into play. While working towards his Ph.D. in 1931, Emmons was fortunate to have collaborated with and trained alongside Benham at Columbia University (M. Silva-Hutner, personal communication, November 20, 1993). As a talented, most thorough, and dedicated young scientist, Emmons earned the respect of Hopkins and Carroll Dodge. In 1929, with his new Rockefeller grant in hand, Hopkins hired Emmons as an associate in mycology under Benham (Campbell, 1983).

Jobs were scarce when Emmons received his Ph.D. in 1935, and Hopkins offered Emmons a position with Arturo L. Carrión at the School of Tropical Medicine in San Juan, Puerto Rico. There, Hopkins saw another opportunity for the young scholar to extend his work on fungal pathogens causing disease in humans and other animals. He was not disappointed, as several notable findings (see scientific contributions) were made and a long and productive association between Emmons and Carrión developed (C. Campbell, personal communication, April 1, 1993; M. Silva-Hutner, personal communication, November 20, 1993).

In 1936, the budget of President Franklin Delano Roosevelt authorized a medical mycologist at the new National Institutes of Health (NIH) in Washington, DC; NIH was moved to Bethesda, Maryland in 1939. With the full backing of Carroll Dodge and Hopkins, Emmons was chosen and became the first medical mycologist hired by the United States government, thus beginning a distinguished career of 30 years. His numerous research accomplishments inevitably led to the demand for his teaching and training expertise. In addition to recruiting and holding capable, practical laboratory assistants, Emmons was a trainer and mentor for a number of well-known medical mycologists. Included in this group were Donald Louria, Ned Fedder, Vincent Andriole, John P. Utz, Herbert Hasenclever, H. Jean Shadomy, John E. Bennett, Kyung Joo Kwon-Chung, Arden Howell, Jr., Samuel B. Salvin, Leo Pine, and Dexter Howard ("Training Trees", Appendix B).

Beginning in 1942, Emmons taught second-year medical students at George Washington University, a post he enjoyed for 20 years as he imparted his knowledge and expertise to numerous health professionals. Emmons was a quiet, methodical teacher and trainer; often, his contributions and insights were only recognized years later (Kwon-Chung and Campbell, 1986). When the Clinical Center opened at the NIH in 1953, Emmons taught medical mycology to all young physicians doing their postdoctoral fellowships at the NIH Clinical Center, especially those doing research in systemic mycoses. (J. Utz, personal communication, January 7, 1994). Appropriately, he was the key figure most often cited in the questionnaire for his contributions to medical mycology (40 responses, Appendix A). Emmons believed his major contribution was showing how prevalent mycoses were and that their etiologic agents could be found almost anywhere. He said this to Michael McGinnis shortly before his death (M. McGinnis, personal communication, July 22, 1993).

Duke University

Norman F. Conant was one of William Weston's students at Harvard University, earning his Ph.D. in 1933. Conant described himself in 1931 as "just another graduate student under Weston, for there were few centers interested in medical mycology in the country at that time, and Harvard was not included in that list" (Conant, 1969). To support his family during the Great Depression (1929–1940), Conant accepted a Traveling Fellowship from Harvard to study at the Institut Pasteur in Paris with the well known mycologists Maurice Langeron and Guerra. During that time Langeron taught a course in medical mycology, part of which required spending some time with the famous Raimond Jacques Sabouraud (Campbell, 1984). Shortly after Conant's return to Boston as a research assistant at the Massachusetts General Hospital, David T. Smith invited him in 1935 to interview at Duke University for the position of Instructor in Mycology in the School of Medicine, and mycologist for the University Hospital (Mitchell, 1991). Conant was offered the position, thus becoming the first individual hired as a medical mycologist at a medical school.

Conant thrived in the clinical and research atmosphere encouraged by D.T. Smith. Mostly self-taught as a medical mycologist through a combination of curiosity and circumstance, Conant, who was kept out of medical school by the Great Depression, was ready when the outbreak of World War II severely challenged the medical community. In 1943, Conant enrolled for a three-month course at the Army Medical School in Washington, DC, primarily to study entomology and malariology in order to serve as a substitute for regular instructors who were being called into active duty. This course was Conant's only formal instruction in medical microbiology. But, during the course he was disturbed by the lack of knowledge and adequate instruction in medical mycology. Conant stepped forward and tried to inform the members of his class, an action quickly noticed by the Director of the course. As a result, he was invited to become the instructor in medical mycology for these three month sessions, a relationship that continued until 1946 (Campbell, 1984).

One of Conant's most important contributions came when the Army Surgeon General asked him to write a textbook on medical mycology. Military training and combat had exposed large numbers of troops to fungal diseases, and the escalating incidence of fungal infections made medical mycology a necessary discipline that demanded the availability of a textbook. Conant's book, *The Manual of Clinical Mycology*, was written in collaboration with David Tillerson Smith, Daniel Stover Martin, Roger D. Baker, and Jasper Lamar Callaway, all physicians from Duke University (Appendix C). It was an instant success when it was first published in 1944 and became a benchmark for all textbooks in the field well into the 1980s (Questionnaire, 14 responses, Table 3, Appendix A).

Following the war, the demand for medical mycology training increased, and Conant soon realized that he would have to offer a formal course. The interest from practitioners, especially physicians, microbiologists, and medical technicians, led to Conant's first summer course in 1947. He developed the course for physicians, but it was open to all who were interested in medical mycology. The course soon acquired a world-wide reputation and was taught by Conant until 1973, interestingly, to the same type and number of students each year (Mitchell, 1991). For four weeks in June and early July, Conant immersed his students in the practical aspects of correctly isolating and identifying fungi pathogenic to humans (Mitchell, 1991). With the exception of updates on advancements in the field, the content and format of the course never varied greatly. The first two hours of each day were dedicated to lectures (Conant and other faculty) and discussion, followed by laboratory studies of cultures and other demonstration materials. The afternoons were open for reading, research, additional laboratory work as needed, and the identification of his "infamous unknowns". No student received a completion certificate until a prompt identification was made of all the pathogenic fungi and many of the contaminants they would likely encounter in their future work. These free sessions sparked keen competition and camaraderie, resulting in a healthy learning environment (Campbell, 1984; Mitchell, 1991).

Conant directed approximately two dozen dissertations and supervised many other students from around the world. Outstanding leaders and trainers in medical mycology from Conant's program included Leanor Haley who went to the the CDC, Lorraine Friedman to Tulane University in New Orleans, Carlyn Halde to the University of California in San Francisco, and George Cozad to the University of Oklahoma in Norman ("Training Trees", Appendix B).

University of Oklahoma, Norman and Oklahoma City

A key figure in medical mycology training in the midwest was Howard W. Larsh, who was Chairman of the Department of Botany and Microbiology at the University of Oklahoma, Norman until 1976. He had worked with the federal government during the war, and shortly afterwards in 1946, he studied medical mycology with Norman Conant and his colleagues at Duke. Larsh returned to the University of Oklahoma, where in 1947 he established M.S. and Ph.D. programs in medical mycology. He was a teacher and mentor for over 80 graduate students and postdoctoral fellows during his career ("Training Trees", Appendix B). Although Larsh set high academic and professional standards for himself and his students, he always had time to help them. Medical mycology was a favorite topic to Larsh, who had a way of letting his students know that they were special to him. Norman Goodman and Glenn Roberts (1994) remember that, "When he attended professional meetings with his students, you would find him with them instead of with his peer group" (p. 2).

Larsh trained many important contributors to medical mycology, including several who established important training centers that continued strong programs into the 1990s ("Training Trees", Appendix B). He was a leading advocate for more training, having worked closely with the pathogenic fungi endemic in the southwest and with Michael Furcolow in the Kansas City Field Office of the United States Public Health Service during the summer (1950 to 1962) and from September, 1952 to June, 1953. During Larsh's tenure at Oklahoma, the University had training grants from NIH (G. Cozad, personal communication, January 11, 1994). In a 1971 statement to a Research Corporation committee responsible for the administration of the Brown-Hazen program, Larsh stated: "It is not at all uncommon for a graduate registered technician to remark that he was not permitted to examine fungus cultures, as they might be pathogens" (Baldwin, 1981, p. 135.)

The Centers for Disease Control (CDC)

Any discussion of medical mycology training must include the contributions of Libero Ajello and his team at the Centers for Disease Control (CDC) in Atlanta. During 43 years of service, Ajello became renowned as a teacher, researcher, and consultant. He was the perfect disciple to carry forth the mission of CDC to promulgate the best medical mycologists had to offer to prevent and eradicate fungal diseases. Ajello was mostly a self-taught medical mycologist having obtained his Ph.D. in general mycology in 1947 under John Karling, Head of the Mycology Department at Columbia University in New York (Chezzi, 1992). In 1943, Ajello was hired by the United States Office of Scientific Research and Development (OSRD) to work on tinea pedis among Army recruits at Fort Benning, Georgia. He had the opportunity of a few weeks of training under Rhoda Benham before he went to Fort Benning as part of a team headed by Hopkins from Columbia University. Ajello followed this work with another two years with the OSRD at the Johns Hopkins University where he conducted drug research on tinea pedis (L. Ajello, personal communication, January 4, 1994).

After graduating from Columbia University's Graduate School of Arts and Sciences in 1946, Ajello won a three-day national examination for a medical mycologist staff position at the newly established CDC Mycology Unit. During Ajello's two months at Duke University, he worked in Conant's diagnostic laboratory identifying fungi and expanded the knowledge that he had acquired while doing research on dermatophytes for the Armed Forces during World War II. Athlete's foot, a common infection among service men due to poor hygiene in the field and a lack of proper bedding and facilities, had also become a problem for the Army during World War II.

At the CDC, Ajello developed intensive four-week and two-week courses that were given world-wide. The four-week course covered the identification and taxonomy of all known fungi pathogenic to humans, including serology and histopathology. The two-week course covered both cutaneous and subcutaneous infections (one week) and systemic mycoses (one week). The courses were well designed and illustrated, and included a teaching and reference manual (L. Ajello, personal communication, January 4, 199; W. Kaplan, personal communication, January 6, 1994).

In addition, Ajello, as an Adjunct Associate Professor at Emory University, and his CDC colleagues at the invitation of Professor Morris Tager, Chairman of the Microbiology Department, presented lectures and laboratory sessions for the second year medical students at Emory's School of Medicine during several decades. Under the auspices of the World Health Organization, the Pan American Health Organization, and various Ministries of Health, workshops on the mycoses were presented in Burma, Brazil, Colombia, Costa Rica, Ecuador, Egypt, Guatemala, Honduras, Israel, Italy, Jamaica, Thailand, Venezuela, and other countries. From 1979–1990, Ajello served as Principal Investigator and Head of the World Health Organization's Collaborating Center for Mycotic Diseases at the CDC.

Other important medical mycologists at the CDC, who were integral members of Ajello's training and research teams included Lucille K. Georg, Morris Gordon, William Kaplan, Leo Kaufman, Leo Pine, Paul Standard, and Arvind A. Padhye ("Training Trees", Appendix B). The work of these well-known investigators is examined in the research section of Chapter IV.

However, "the formative years" of medical mycology were drawing to a close, and the end of World War II brought sweeping changes to most areas of life in the United States with medical mycology being no exception. As the discipline found itself poised on the brink of new directions, the thoughts of Morris Gordon (1993) may sum up the era the best. In 1946, when searching for the best program to pursue his doctoral studies with a concentration in medical mycology, Gordon found only two programs in the United States offering a formal curriculum: Columbia University with Rhoda Benham or Duke University with Norman Conant. Gordon chose Duke and trained under F. Wolf in botany.

Gordon had been advised by Conant to major in botany with specialization in medical mycology. Upon arriving, Gordon found himself studying with Leanor Haley, Lorraine Friedman, and other students under the tutelage of Conant (Gordon, 1993). Thus, by the end of "the formative years" many of the leaders of the new era were in training or initiating their programs at centers across the United States, including the CDC, Oklahoma, Michigan State University, and Tulane University, in addition to Columbia University, Duke University, UCLA, and the NIH.

Mycology books

In the beginning of "the formative years", several books were published that served as important contributions to medical mycology (Appendix C). Bacteriologists were among the first investigators to search for tools to improve the laboratory diagnosis of pathogenic fungi. The meager information available to these early pioneers in medical mycology had to be gleaned from the various classical, medical, and bacteriology textbooks of that time. *Fungous Diseases: A Clinico-Mycological Text* published in 1932 by Harry P. Jacobson was the first book in the United States to deal solely with medically important fungi. It was strictly descriptive from a clinical standpoint without the etiologic agents being discussed (Kobayashi, 1993). Closely following this book was the publication of *Medical Mycology: Fungous Diseases of Men and Other Mammals* by Carroll W. Dodge in 1935 (Appendix C). Dodge's book became very important, for it provided the first complete descriptions of all the fungi known or considered at that time to be pathogenic to humans and other animals (M. McGinnis, personal communication, March 17, 1994; Kobayashi, 1993).

IV. The advent of antifungal and immunosuppressive therapies: 1950 to 1969

Figure 51. Lucille K. Georg (1st, second row) (p. 60) , Gordon Morris (4th, 3rd row), and Ajello (6th, 3rd row) (Circa 1950s) (Courtesy of Dr. Gordon)

Figure 52. Lucille Georg (pp. 60, 65, 71, 73, 77) (Courtesy of Dr. D. MacKenzie)

Figure 53. Lucille K. Georg (seated), Libero Ajello, Leo Kaufman, and William Kaplan at CDC (pp. 60, 77) (Courtesy of Dr. Kaufman)

Figure 54. Margarita Silva-Hutner (first row, 2nd) (pp. 60, 73-74), Irene Weitzman (first row, 3rd) (p. 71), William Merz (top row, 2nd) and Gladys Rebatta (second row, 6th) (Courtesy of Dr. Silva-Hutner)

Figure 55a. Lorraine Friedman (pp. 60, 68, 75) (Courtesy of Dr. Sinski)

Figure 55. Morris F. Shaffer (pp. 60, 75) (Courtesy of Dr. Restrepo)

Figure 56. Dexter Howard (pp. 60, 71, 72, 76)

52

ISHAM MEETING 1967
NEW ORLEANS

Figure 57. Juneann Murphy (second row, 2nd), George Cozad (second row, 3rd) (p. 60), and seated, Y. Al Doory, Ernest Chick, Norman Goodman, Mrs. Goodman, Charles Smith, and Howard Larsh (p. 76) (Courtesy of Goodman)

Figure 59. John P. Utz (pp. 63, 68, 70, 74-77)

Figure 58. Rachel Fuller Brown (1898-1980) and Elizabeth Lee Hazen (1885-1975) (pp. 62-63) (Courtesy of Dr. Gordon)

Figure 60. Morris Gordon (pp. 64-66, 68, 70), Lorraine Friedman, and J. Donald Schneidau (pp. 68, 75) (Courtesy of Dr. Gordon)

Figure 61. Leanora Haley (1st row, 6th) (pp. 65, 77) and her mentees: Timothy Cleary (second row, 10th), Michael McGinnis (second row, 11th) (Courtesy of Dr. Cleary)

Figure 62. Smith Shadomy (pp. 66, 68, 70) and H. Jean Shadomy (pp. 68, 71) (Courtesy of Dr. Sinski)

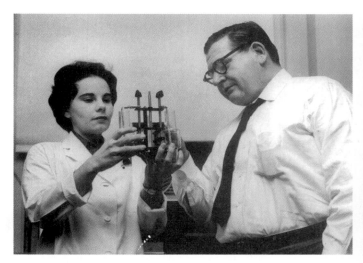

Figure 63. Walter Nickerson (pp. 66-67) (Courtesy of Dr. Reiss)

Figure 64. John Bennett (pp. 68, 70, 75)

Figure 65. William Winn (1903-1967) (p. 69) (From: Huntington, R. W. 1985. Four great coccidioidomycologists. J. Med. Mycology. 23, p. 361-370)

Figure 66. Michael Furcolow (1907-1985) (p. 69) (From: MMSA Archives)

Figure 67. David Taplin (pp. 70-71) (Courtesy of Dr. MacKenzie)

Figure 68. Gerbert Rebell (pp. 70-71) (Courtesy of Dr. Sinski)

56

Figure 69. Kenneth Raper (2nd) and Kyung Joo Kwon-Chung (4th), his mentee) (p. 71, 75) (Courtesy of Dr. Kwon-Chung)

Figure 70. J. Fred Denton (right) and Arthur DiSalvo his mentee (p. 71) (Courtesy of Dr. DiSalvo)

Figure 71. Eugene McDonough (p. 71) (Courtesy of Dr. Sinski)

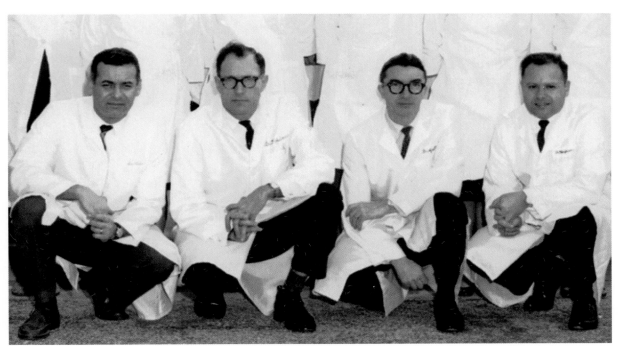

Figure 72. Leo Pine (p. 72), William Kaplan, Libero Ajello, and Leo Kaufman at CDC in the 1960s (p. 77) (Courtesy of Dr. Kaufman)

Figure 73. Everett S. Beneke and his mentee Alvin L. Rogers (p. 72, 74 (Courtesy of Dr. Sinski)

Figure 74. Glenn Bulmer (p. 72)

58

Figure 75. John Rippon (p. 73)

Figure 76. John Rippon (first row) (p. 73), Nina Dabrowa (second row, 3rd), Libero Ajello and Lorraine Friedman (third row, 2nd and last), Arvind Padhye (fourth row, 2nd), Charlotte Campbell, and Alvin Rogers (top row, 8th and last) (1970) (p. 72, 74) (Courtesy of Dr. Rippon)

Figure 77. Arturo Carrión and his mentee Margarita Silva-Hutnert at Columbia in 1949 (p. 73) (Courtesy of Dr. Silva-Hutner)

Figure 78. Stanley Rosenthal (6th, first row) (p. 74) and his mentor VanBreuseghem (4th, first row) (Courtesy of Dr. Rosenthal)

Figure 79. Dexter Howard and his students (Betty Wu-Hsieh and Nina Dabrowa) (p. 76) (Courtesy of Dr. Howard)

Figure 80. Everett S. Beneke and Jim T. Sinski (p. 78) (Courtesy of Dr. Sinski)

Overview: 1950 to 1959

Drastic changes were needed in medical mycology by the end of 1949, because the mortality rate for patients having invasive mycoses ranged between 50 and 100%, depending on the severity of the disease. The 1950s experienced the discovery of nystatin, the first active antifungal drug to be used in humans, followed by the development of stilbamidine and, more importantly, amphotericin B. John Fisher and John Perfect stated that although the azoles have been important new antifungal agents, the use of amphotericin B opened the field of antifungal therapy and remains the gold standard (J. Perfect; J. Fisher, questionnaires). The development of antifungal agents is ranked as the second most important event for the development of medical mycology in the United States (15 to 24.7% responses to the questionnaire, Tables 3 and 4, Appendix A), because it stimulated an intense interest among both the medical and pharmaceutical communities, enhanced public awareness, and legitimized the discipline as a distinctive field in medicine. Meanwhile, royalties from the sale of nystatin provided much needed financial support for training and research distributed through the Brown-Hazen Committee.

Other important contributions made during this decade included serological tests for the diagnosis of coccidioidomycosis and histoplasmosis, and practical nutritional tests for the identification of selected dermatophytes were developed. The antigenic composition of the polysaccharide capsule of *C. neoformans* was determined, and the detection of fungi in tissues was improved. The relationship of opportunistic fungal infections to antibiotic therapy, the use of cytotoxic or immunosuppressive therapy, certain malignancies, and diabetes were partially clarified. The preliminary transition from studies of the biology of fungi to those at the cellular level were made between 1953 and 1959, mostly by scientists at Rutgers University and the NIH.

In the 1950s, Lucille K. Georg, William Kaplan, and Leo Kaufman arrived at the CDC. Margarita Silva-Hutner replaced Georg at Columbia University and became the new Director of the Laboratory in 1956, shortly before Rhoda Benham's death on January 17, 1957. Early in 1955, the physician Morris F. Shaffer, who had arrived at Tulane University in 1943, traveled to Berkeley and recruited Lorraine Friedman to develop a medical mycology unit at Tulane University (Shaffer, 1985).

The Veterans Administration Armed Forces Cooperative Study Group on Coccidioidomycosis was formed in 1955. This group, currently known as the Coccidioidomycosis Study Group, has met annually to discuss the state of coccidioidomycosis research. In August 1994, they convened an international conference at Stanford University to coincide with the centennial of the description of the first two American cases of coccidioidomycosis at Stanford University by Emmet Rixford and Thomas Caspar Gilchrist in 1896. The VA Armed Forces Study Group also sponsored other mycoses cooperative studies from which much scientific information was published in the 1960s.

Dexter Howard began his work at the University of California at Los Angeles in 1956 as an Instructor of Mycology. He is recognized for his numerous basic research contributions on dimorphism and host-parasite interactions in histoplasmosis capsulati, as well as for the training of numerous masters and doctoral students ("Training Trees", Appendix B). In 1957, George Cozad came to the University of Oklahoma as a Research Associate in medical mycology under Howard Larsh. In 1958, Charles D. Jeffries began his medical mycology career at Wayne State University. In 1959, Hillel Levine was appointed Chief of the Mycology Division at the Naval Biological Laboratory in Oakland, California. Trials of new antifungals, in vitro assays, animal models, development and evaluation of coccidioidin and spherulin skin test reagents, and laboratory safety standards were among the important activities performed at that institution.

1950 to 1959

Immunocompromised patients versus opportunistic infections

The association of a high incidence of severe fungal infections with antibiotic and steroid therapy, drug addiction, and leukemia or lymphoma, especially Hodgkin's disease was first documented in the 1950s. Lorenz Zimmerman (1950) reported three cases of *Candida* and *Aspergillus* spp. endocarditis "apparently stimulated" by the constant administration of penicillin at the Walter Reed General Hospital. At Montefiore Hospital (New York City),

Richard Torack (1957) reported that 13 fungal diseases were found at autopsy in patients who had been treated with either antibiotics, steroids, or both. An increase from three cases (between 1933 and 1952) to 13 cases (between 1952 and 1956) of fungal corneal infections and post-traumatic vitreous abscesses was documented at the Registry of Ophthalmic Pathology. These 13 patients had received topical, steroid-bacterial antibiotic combinations (Haggerty and Zimmerman, 1958).

Since 1894, various individuals believed that cancer was caused by a yeast. However, in 1902 E.H. Nichols had concluded that there was no evidence that yeasts caused human cancer. In 1954, Lorenz Zimmerman and Henry Rappaport evaluated the divergent views since 1932 concerning the controversial association between cryptococcal infections and malignant lymphoma, particularly Hodgkin's disease. At the Armed Forces Institutes of Pathology (Washington, DC), they reviewed 60 *C. neoformans* cases and grouped them according to the extent of the infection. They found that dissemination occurred more commonly in patients with malignant lymphoma or leukemia (18 of 24 disseminated cases), and clearly negated the prevalent, but erroneous, concept that *C. neoformans* was responsible for Hodgkin's disease.

John Keye and W. Edwin Magee (1956) at Washington University surveyed clinical records, autopsy data, gross and histologic material, and microbiological data on 88 cases of mycoses among 15 845 autopsies performed between 1919 to 1955. Although the incidence of primary fungal infections had not increased after 1947, there was a striking increase in the incidence of secondary fungal infections. In general, cryptococcosis and zygomycosis occurred in patients who had lymphomas or leukemia; histoplasmosis capsulati, candidiasis, and aspergillosis complicated other diseases as well. The frequent use of multiple antibiotics, cytotoxic chemotherapy for leukemia and lymphoma, and cortisone contributed to the increased incidence of fungal infections observed after 1947 (Keye and W. Magee, 1956). Although most of these cases were diagnosed based upon the morphology of the fungus in tissue, these reports alerted the scientific community to the fact that various immunologic and nonimmunologic factors could predispose individuals to opportunistic fungal infections.

During the 1950s, attempts were made to understand the mechanisms by which diabetes, antibiotics, and cortisone increased the susceptibility of the host to fungal infections. Heinz Bauer, John Flanagan, and Walter Sheldon (1955), who were at Emory University, experimentally reproduced the lesions of cerebral zygomycosis (as mucormycosis) in diabetic and nondiabetic rabbits. Because the animals were infected by intranasal instillation of sporangiospore suspensions, the infection began in the nasal mucosa and rapidly extended to adjacent tissue, bone, and finally the brain. In contrast, the nondiabetic rabbits had fewer and smaller lesions and the lesions were surrounded by many neutrophils, which did not show the same degenerative changes seen in the diabetic rabbits. Herschel Sidransky and Lorraine Friedman (1959), at Tulane University, determined that the combination of antibiotics plus cortisone or cortisone alone rendered animals, that had inhaled *Aspergillus flavus* conidia, highly susceptible to developing fatal pulmonary infections.

Cerebral phaeohyphomycosis

Chapman Binford, R.K. Thompson, Mary Gorham, and Chester Emmons (1952) recovered *Cladophialophora bantiana* (as *Cladosporium trichoides*) from a brain abscess of a 22-year-old black male, who had complained of frontal headaches for two weeks, and by the time of hospital admission, was sleepy and drowsy. Microscopically, the central part of the CNS clinical specimen contained "degenerating purulent exudate in which were found numerous light brown filamentous structures" (Binford et al., 1952, p. 536). The hyphae measured 1 to 2 μm in diameter, exhibited branching and contained numerous septations. Gray-to-brown-to-olive black colonies were grown, which contained phaeoid hyphae and "2 μm in diameter conidiophores, often septate, straight to flexuous, bearing long sparsely branched chains of conidia" (Binford et al., 1952, p. 541). Emmons proposed the name *Cladosporium trichoides*, signifying its hyphal growth in human brain tissue. A new species was added to the list of fungi that could cause opportunistic infections in humans. This fungus was transferred to the genus *Xylohypha* in 1986 by Michael McGinnis, Dante Borelli, Arvind Padhye, and Libero Ajello, and to the genus *Cladophialophora* in 1995 by G.S. de Hoog et al.

Dermatophytosis

The dermatologist Albert Kligman, from the University of Pennsylvania, declared in 1952 that, "little basic knowledge has accumulated on ringworm of the scalp" (p. 231) since Raimond Sabouraud's *Les Teignes* (1910). Epidemics of tinea capitis caused by *Microsporum audouinii* in the early 1950's prompted important, although currently ethically questionable investigations. This epidemic, which was known as the Atlantic City School Board Epidemic, was so wide spread that infected schoolchildren were required to wear hats in order to go to school (M. Silva-Hutner, personal communication, November 20, 1993). Kligman published two papers in 1952 and 1955 on the pathogenicity of *M. audouinii* and *M. canis*. His data were derived from observations on experimentally infected humans that were carried out in a state institution for congenital "mental defectives", where tinea capitis was prevalent. He found that only 60% of children and adults were susceptible to experimental infection. He divided the natural course of the infection into four phases: (i) first few days incubation, (ii) three month enlargement of original lesions and development of new ones, (iii) refractory or static infection, and (iv) involution. Tinea capitis comprised two types of infections, scalp and hair (Kligman, 1952). Hyphae were observed to grow downward into the hair follicle and break up into chains of cells ("primary spore formation") during the incubation period. During the enlargement period, new follicles were infected and external branches of the intrapilary hyphae segmented into short chains of ectothrix conidia. These conidia ceased to be formed during the involution phase (Kligman, 1955).

Portal of entry for Coccidioides immitis and Blastomyces dermatitidis

Walter Wilson, Charles E. Smith, and Orda Plunkett (1953) had the opportunity to follow the course of a case of coccidioidomycosis involving the skin of a patient seen at the University of California, Los Angeles. The patient was an embalmer who had severely abraded the skin of his finger against a casket. During the embalming of a body, bloody and seropurulent material from the visceral organs containing *C. immitis* cells had come in contact with the abraded skin. From the clinical observations and the mycologic data, it was concluded that "the disease originates very rarely as the result of primary cutaneous inoculation" (Wilson, Smith, and Plunkett, 1953, p. 233). Skin lesions are nearly always the result of dissemination of *C. immitis* to the skin by hematogenous means. A similar conclusion was reached concerning blastomycosis by Jan Schwarz and Gerald Baum (1951) who reviewed 154 cases of blastomycosis, 58 from their institution (Cincinnati General Hospital). Their conclusion was based on the following observations: skin lesions followed involvement of the lung; rapid development of skin lesions was associated with simultaneous appearance of articular or bone lesions; diseases produced by inoculation almost always began on the feet or hands; and lesions started as closed eruptions. Thus, it was established that blastomycosis and coccidioidomycosis initiated in the respiratory tract.

Antifungal therapy

During the 1940s, awareness that fungal diseases were common and prevalent emphasized the need for new antifungal agents. Toxicity and other undesirable pharmocologic properties of earlier compounds made them unsuitable for use in humans. A deliberate search for an antifungal agent was instigated by Gilbert Dalldorf at the New York State Department of Health to treat the large numbers of servicemen and women returning from the tropics with severe fungal infections. A major contribution to the development of medical mycology occurred when Elizabeth Lee Hazen and Rachel Fuller Brown (1950) announced from this institution the discovery of nystatin (as fungicidin, named for New York State) at the autumn meeting of the National Academy of Sciences in Schenectady, New York. Hazen, a bacteriologist, and Brown, a chemist, as a team were able to identify, characterize, and purify nystatin following its detection in cultures of *Streptomyces noursei* (named after William Nourse, the farm's owner) near the barn of a dairy farm in Fauquier County, Virginia (MMSA Newsletter, June 1972). The screening of soil samples for antifungal actinomycetes was conducted by a selective plating technique in which *C. albicans* and *C. neoformans* served as the screen fungi. Brown and Hazen also demonstrated the in vitro and in vivo activity of nystatin in laboratory (experimental) animal studies.

Brown and Hazen donated the royalties of their discovery, which amounted to more than $13 000 000 by 1976 when the patent expired, to a special fund designated the Brown–Hazen Fund. These funds were used for advancing the medical sciences in general and in the 1970s, exclusively for medical mycology. Nystatin was patented and made available to E.R. Squibb and Sons by the Research Corporation in New York City. Hazen and Brown were inducted posthumously into the National Inventors Hall of Fame. They were only the second and third women to be so honored (M. Gordon, personal communication, January 10, 1994; Research Corporation News, January 6, 1994).

Until 1950, blastomycosis was treated with iodides, X-ray therapy, and surgical procedures. These approaches only provided temporary remission of the cutaneous lesions, but a 92% case fatality was expected in patients with disseminated *B. dermatitidis* infections. Following W.O. Elson's (1945) report that propamidine was effective in vitro against *B. dermatitidis*, Emanuel Schoenbach, Joseph Miller, Milton Ginsberg, and Perrin Long (1951) treated three patients with propamidine and a related compound, the aromatic diamidine, stilbamidine. These patients who had systemic blastomycosis showed improvement, but unfortunately the toxicity of stilbamidine necessitated a brief and intermittent therapy. A derivative of stilbamidine, 2-hydroxystilbamidine, was given to another three patients with blastomycosis for a prolonged and extensive course without evidence of toxicity (Snapper and McVay, 1953).

In the meantime, the search for more effective antifungal agents continued at the E.R. Squibb Institute. W. Gold, H.A. Stout, J.F. Pagano, and R. Donovick (1955 to 1956) isolated the two antifungal "materials" amphotericin A and B, from a broth culture of *Streptomyces nodosus*. In vitro broth dilution assays demonstrated that B was more active than A, but less soluble, against the yeasts and filamentous fungi that they had tested. Both A and B preparations demonstrated antifungal activity in vivo when administered subcutaneously to experimentally infected animals. Fungal cultures and consultations for their study were provided by Margarita Silva-Hutner at Columbia University and Elizabeth Hazen at the New York State Health Department (Steinberg, Jambor, and Suydam, 1955 to 1956; Silva-Hutner, personal communication, November 20, 1993). Bernard Steinberg et al. (1955 to 1956) demonstrated that the oral administration was not as effective as the intravenous doses. In 1957, Richard Harrell and Arthur Curtis at the University of Michigan treated four "stilbamidine-failure male patients" successfully with intravenous amphotericin B. These four patients were given 50 mg of amphotericin B that was suspended in 1000 ml of 5% aqueous dextrose for six to eight hours. A few months earlier, M.J. Fiese (1957) reported a case of disseminated coccidioidomycosis that he had successfully treated orally.

Beginning in 1956, investigators at the NIH studied the in vitro and in vivo efficacy of amphotericin B in experimental cryptococcosis and histoplasmosis capsulati (Louria, Feder, and Emmons, 1956-1957), as well as the efficacy, safety and pharmacokinetics of oral and intravenous therapies given to patients with severe fungal infections (Utz, Louria, Feder, Emmons, and McCullough, 1957–1958; Louria, 1958). Their data indicated that oral or intraperitoneal drug was protective against usually fatal mouse infections by *C. neoforman* and *H. capsulatum* var. *capsulatum*, but the oral formulation was poorly absorbed and only low serum and CSF levels were obtained. In contrast, the intravenous solution produced better blood levels and was slowly excreted; serum concentrations were fungistatic for prolonged periods after the medication was discontinued. Donald Louria (1958) concluded that the intravenous dosages were beneficial for systemic infections when they did not involve the central nervous system.

John Seabury and Harry Dascomb (1958); John Utz, Albert Treger, Norman McCullough, and Chester Emmons (1958–1959); M.L. Littman and P.L. Horowitz (1958); and Littman (1959) used a new formulation of amphotericin B (E.R. Squibb and Sons), which had became available in December 1956 under the trade name Fungizone. The drug was combined with approximately 33 mg of sodium deoxycholate per 50 mg of drug. When suspended, "it formed a golden fluid which had the gross characteristic of a solution even though it was actually a colloidal suspension" (Seabury and Dascomb, 1958, p. 961). Results of their studies indicated that the intravenous preparation was very effective against *B. dermatitidis*, *H. capsulatum* var. *capsulatum*, and *S. schenckii*. Unfortunately, the data were inconclusive for patients with disseminated *C. immitis* infections and *C. neoformans* meningitis. Also, toxic effects were reported that disappeared after therapy. The need for randomized studies and prolonged follow-up was deemed necessary to further evaluate the usefulness of this agent.

Until 1958, studies were directed towards its effect on pathogenic yeasts and dimorphic fungi. In 1958, Ernest Chick, John Evans, and Roger Baker at Duke University determined the effect of this drug in an experimental rat model using the opportunistic zygomycete, *Rhizopus arrhizus* (as *R. oryzae*). They concluded that amphotericin B appeared promising as a therapeutic agent for human zygomycosis.

A third antifungal agent, griseofulvin, was developed in 1956 in Scotland. In the course of a search for new chemotherapeutic "remedies" for the treatment of dermaphytoses, James C. Gentles (1958) observed "that griseofulvin, a metabolite produced by *Penicillium janczewskii* (as *P. griseofulvum*) (Oxford, Raistrick, Simonart, 1939) was effective against *Trichophyton mentagrophytes* infections in guinea pigs" (Martin, 1958, p. 1232). This was the first orally administered antifungal compound against a dermatophyte.

Laboratory diagnostic methods

During the late 1930s, Charles E. Smith and coworkers at the University of California, Berkeley, attempted to neutralize the ability of coccidioidin to elicit a skin reaction. In doing so, they discovered that a mixture of coccidioidin with serum from infected guinea pigs yielded a precipitate. Smith collaborated with two serologists, Karl Meyer and Bernice Eddie, on the CF test. They performed 21 000 precipitin and CF tests for coccidioidomycosis from 1940 to 1949, while Smith was at Stanford University (Smith et al., 1950). Their analysis of the data demonstrated that the combined CF and precipitin tests detected more than 90% of clinically apparent cases of coccidioidomycosis, less than 10% of asymptomatic cases, 60% of pulmonary cavitation lesions, and 99% of disseminated cases. Each test alone detected from 22% (CF) to 44% (precipitin) of these cases. "The tests were specific with respect to viral, rickettsial, spirochetal, bacterial, and most other mycotic infections" (Smith et al., 1950, p. 19). However, there were cross-reactions with *H. capsulatum* var. *capsulatum*. From the results of both skin reactions and serology, it was discovered that the skin reaction is positive first, followed by the detection of precipitins and CF antibodies in primary noninvasive disease. The CF antibodies persisted and an increase in titer corresponded to the severity of the infection; a decrease correlated with patient improvement. These tests became an excellence guide for following and anticipating the course of this infection.

Albert Coons and Melvin H. Kaplan (1950), who were at Harvard Medical School, developed a specific histochemical method for detecting antigenic material in tissues by using fluorescein labeled antibodies. This method was applied initially to demonstrate the capsules of bacteria and later by Warren Eveland and coworkers (1957) for the study of *C. neoformans* in tissues maintained at the Armed Forces Institute of Pathology (Washington, DC). Ralph Vogel and Joseph Padula (1958) and Vogel (1966) at Emory University School of Medicine used the fluorescent stain method (FA) to demonstrate antibodies in sera from patients who had various systemic diseases. Morris Gordon (1958) applied this method to the differentiation of yeasts in the clinical laboratory. The FA technique was used extensively by William Kaplan and Leo Kaufman at the CDC as a reference method to assist in the diagnosis of mycotic infections using tissue.

Special staining techniques for the specific demonstration of fungi in tissues were developed between 1951 to 1953. Examination of tissue sections was difficult because most pathologists did not have sufficient mycologic experience to adequately recognize fungi in tissues. A selective stain for fungi was needed, and the adaptation of Hotchkiss and McManus' stains (1948) provided a solution to the problem. The procedure was based on the fact that the cell wall of fungi is composed of chitin that is absent in higher animals. The Hotchkiss–McManus technique was used to stain the carbohydrates in the fungal cell wall. However, the fungal wall had to be hydrolyzed with periodic acid to release its aldehydes, which then combined with the Schiff reagent to "bring about its decoloration in various shades of red and magenta" (Kligman, Mescon, and De Lamater, 1951, p. 86). A light green counterstain for the tissue provided a good contrast to the deep rose-colored fungi. Since then, the periodic acid-Schiff stain (PAS) has been used routinely in pathology laboratories to detect fungi. The PAS stain was modified further by Mary Francis Gridley (1953), who replaced the periodic acid component with chromic acid in order to demonstrate both hyphal and yeast cells in tissue.

Cycloheximide was added to mycologic media resulting in mycobiotic agar or Mycosel depending upon which manufacturer made the medium. This medium was originally used in 1951 for the selective isolation of *C. immitis* by Lucille K. Georg, Libero Ajello, and Morris Gordon. It inhibits the growth of *C. neoformans* as well as most saprophytic fungi when it contains low concentrations of cycloheximide. The usefulness of cycloheximide in media has had a great impact in laboratory diagnosis.

The nutritional requirements of selected dermatophytes continued to be studied by Rhoda Benham (Silva and Benham, 1952; Benham, 1953; Silva and Benham, 1954; Silva, Kesten, and Benham, 1955) and at the CDC, where Georg had joined Ajello. Beatrice Kesten was a dermatologist who did her postdoctoral training in French and German Universities (1928–1930), and later in Benham's laboratory. She became the first female member of the American Dermatology Association (M. Silva-Hutner, personal communication, November 21, 1993). Georg and Laverne Camp (1957) simplified the procedures for studying dermatophytes in the clinical laboratory from a nutritional point of view. After several studies, they reported selective nutritional profiles for several species of *Trichophyton*. Georg's doctoral dissertation, which was supervised by Benham at Columbia University, was the basis of her nutritional research. Georg and Camp's (1957) differential medium formulations have been used to prepare seven commercially available *Trichophyton* agars (Difco Laboratories, Detroit, MI). These agars have been an invaluable laboratory diagnostic tool. Leanor Haley stated in 1993: "the ability to identify dermatophytes by nutritional requirements and morphology made it possible for many laboratories to report out competent data for the physician and the mycologist" (L. Haley, personal communication, December 31, 1993). Georg and Benham facilitated the identification of the dermatophytes, which are, even today, difficult to identify because atypical or poorly sporulating strains are frequently isolated. In recognition of Georg's work, a new species, *Trichophyton georgiae*, was named in her honor in 1964. In addition, the Lucille K. Georg medallion was established as an award to be conferred upon outstanding medical mycologists by the International Society for Human and Animal Mycology (ISHAM).

In the same year (1957), Ajello and Georg further investigated A.M. Davidson and P.H. Gregory's (1934) discovery in Canada of the formation of wedge-shaped perforations caused by organized groups of hyphae in hair filaments exposed to *T. mentagrophytes*. Ajello and Georg (1957) system-atically studied the morphology, pigmentation, and in vitro hair penetration of 40 isolates of *T. mentagrophytes* and *T. rubrum*. They concluded that "the different manner in which these two species attack hair in vitro is a diagnostic aid for their identification when correlated with morphology" (p. 17).

At the University of Arkansas Medical Center, Douglas Heiner (1958), adapted Örjan Ouchterlony's (1949) agar-precipitin technique that was used in bacteriology for the immunologic diagnosis of histoplasmosis capsulati. He noted that CF tests were not advantageous because of cross-reactions and the unsuitability of anticomplementary sera. Heiner (1958) utilized the Ouchterlony technique to test more than 2000 sera from patients with culturally proven blastomycosis, coccidioidomycosis, histoplasmosis capsulati, and patients suspected of having these mycoses, as well as "normal" volunteers and patients with other diseases. Examination of precipitin bands demonstrated that histoplasmin contained at least six distinct soluble antigens. He concluded that sera from proven cases of histoplasmosis capsulati showed what he called "h" and "m" bands; "m" bands could be seen following a histoplasmin skin test; "h" and "m" bands correlated with high CF titers; and the absence of a precipitin band (to histoplasmin) ruled out active histoplasmosis. This easy to perform test has been useful for many years for the presumptive diagnosis of histoplasmosis capsulati.

Taxonomy and classification

A taxonomic issue was clarified by Rhoda Benham in 1950 when she reviewed the different names applied to the yeast *C. neoformans*. The generic name, *Cryptococcus*, was adopted by Jean Paul Vuillemin in 1901 for a group of pathogenic yeast-like fungi that did not develop ascospores. Vuillemin gave the name *C. hominis* to Otto Busse and A. Buschke's European strain, known today as *C. neoformans*, which had been isolated originally in 1894. J. Lodder compared F. Sanfelice's *Saccharomyces neoformans*, which was isolated from a peach, with Busse and Buschke's strain and found them to be identical (Benham, 1950). Therefore, Benham (1950) concluded that the fungus associated by J.L. Stoddard and E.C. Cutler (1916) and others with meningeal infections were actually the same fungus, that is, *Cryptococcus neoformans*. She also stated that although yeast cells are found in tissues in both blastomycosis and cryptococcosis, "blastomycosis is an entirely different disease, caused by a mycelium-producing fungus, *Blastomyces dermatitidis*" (Benham, 1950, p. 1312).

Morris Gordon (1951), while he was at the CDC, regularly attended a skin clinic at Grady Hospital in Atlanta. There, he saw many cases of pityriasis versicolor in which the lesions contained numerous yeast cells and mycelial fragments that were not being successfully cultured. Benham (1939) had cultured *Pityrosporum ovale* from skin by adding unsaturated fatty acids to the culture medium. Gordon (1951) was able to culture "spherical forms" of *Malassezia furfur* (*P. ovale*) by adding olive oil or saturated fatty acids to Sabouraud dextrose agar. Because the "spherical forms" had different morphological and physiological characteristics than those of *P. ovale*, Gordon (1951) named this fungus *P. orbiculare*. Ruth C. Burke (1961) described pityriasis versicolor in individuals experimentally infected with *M. furfur* (as *P. orbiculare*) and two years later Frances Keddie and Smith Shadomy (1963) studied the morphologic and antigenic relationships between *P. orbiculare* and *M. furfur*. M. Gordon then stated that: "Everyone now agrees that *P. ovale* and *P. orbiculare* are forms of *M. furfur*" (M. Gordon, personal communication, January 10, 1994).

Ecology

The prevalence of *C. neoformans* throughout the world suggested that this fungus is a widely distributed saprophyte in nature. Chester Emmons (1951), while he was searching for the *capsulatum* variety of *H. capsulatum* in soil, isolated in mice four strains of *C. neoformans* from soil samples collected in Loudoun County, Virginia. Further studies permitted Emmons (1955) to observe that *C. neoformans* was present in pigeon feces and that such pigeon feces constituted a good medium for the fungus to grow in competition with other soil microorganisms. Emmons stated later that his discovery of *C. neoformans* in old pigeon nests was pure serendipity, since he was not looking for this organism. At that time, Emmons (1955) was not able to appreciate the significance of his findings, because of the rarity of cryptococcosis. The association of *C. neoformans* infections with immunosuppression was recognized later.

In 1951, Libero Ajello and Louis Zeidberg isolated *Pseudallescheria boydii* from soil while they were searching for *H. capsulatum* var. *capsulatum* in Williamson County, Tennessee; Chester Emmons had previously isolated *H. capsulatum* from soil in 1949. Zeidberg, Ajello, Ann Dillon, and Laliah C. Runyon (1952) found that the *capsulatum* var. of *H. capsulatum* was isolated more frequently from soil that was enriched with chicken feces. Thus, chicken coops and chicken yards were shown to be potential point sources of infection for the development of histoplasmosis capsulati.

Immunochemistry

Edward Evans, at the University of Southern California, investigated the antigenic composition of *C. neoformans* between 1949 and 1953. It was difficult to perform serologic investigations of *C. neoformans* because the antibody response in experimentally immunized animals was poor. However, Evans (1949) successfully developed antisera with high agglutinin titers by injecting encapsulated formalin-killed *C. neoformans* cells into rabbits. Evans in 1949 and 1950 and Evans and John Kessel in 1951, reported that "*C. neoformans* may be divided into three serologic types, A, B, and C, on the basis of antigenic differences revealed by agglutination reactions" (Evans, 1950, p. 429). Isolation of the capsular polysaccharide of types A, B, and C provided similar results by precipitin reactions. Further experiments demonstrated that the capsular materials of type B had two carbohydrate fractions called SB1 and SB2, which contained xylose, mannose, and glucose by paper chromatography (Evans and Theriault, 1953). Seventeen years later (1966), Ralph Vogel identified the fourth serotype, D (as R).

Physiology and nutrition

The fact that some important fungal pathogens are dimorphic indicated the importance of studying the phenomenon of their conversion from a mycelium form to the corresponding tissue form. Walter Nickerson (1953), who was at Rutgers University in New Brunswick (New Jersey), initiated basic research studies dealing with the mechanisms of cellular division and morphogenesis of *C. albicans*. Nickerson and Zbigniew Mankowski (1953) found that the replacement of glucose by soluble starch, glycogen, or dextrin diminished the filamentous growth of

C. albicans. The growth of budding yeast cells was dependent on glucose as the source of carbon. Later, Nickerson (1954) was able to identify for the first time a cellular oxidation mechanism at the flavoprotein locus, which was essential for cellular division by budding, but not for growth. He demonstrated this metabolic locus by the development of a filamentous mutant of *C. albicans*. The morphological mutant differed from the parent only in "its impairment to a cellular oxidation mechanism at a flavoprotein locus" (Nickerson, 1954, p. 493). Nickerson was the major professor of Errol Reiss, who later went to work at the CDC, where he became Chief of the Molecular Biology Section of the Division of Bacterial and Mycotic Diseases. In 1959, Gian Kessler and Nickerson isolated "clean cell walls" of three *C. albicans* strains and determined that they contained a glucan-protein and two distinct glucomannan-proteins. Samuel B. Salvin (1952), at NIH's Rocky Mountain Laboratory in Hamilton, Montana, discovered that *C. albicans* was the second pathogenic fungus shown to produce a toxic effect after inoculation into laboratory animals. Swiss mice were killed within 48 h following intraperitoneal injection of a suspension of dead yeast cells.

Two important papers appeared in the literature between 1957 and 1958 concerning the nutritional requirements of the parasitic forms of *C. immitis* and *H. capsulatum* var. *capsulatum*. John Converse (1957), at Fort Detrick, Maryland, grew the spherules of *C. immitis* in a medium composed of glucose, ammonium acetate, and inorganic salts. The addition of an anionic surface active agent to the medium maintained the tissue form through serial transfers. In 1954, studies conducted at the NIH by Leo Pine regarding the conversion of *H. capsulatum* to its yeast form resulted in the development of a three amino-acid medium for yeast conversion (Pine and Peacock, 1958). Yeast conversion was influenced by the growth requirements of the yeast (stimulated by citric acid); growth of the mycelial form was inhibited by the presence of citrate at 30° and 37°C (Pine and Peacock, 1958). Salvin's (1949) finding that sulphydril groups affected by temperature determined the growth forms of *H. capsulatum* was later confirmed by Leo Pine and Carl Peacock (1958). Salvin's (1949) medium and his observations led to further basic discoveries by Gerald Medoff and George Kobayashi at Washington University in St. Louis, Missouri.

Overview: 1960 to 1969

During the 1960s, evaluation of large numbers of patients in cancer, leukemia and transplantation centers alerted the medical community to the fact that fungal infections were an important cause of morbidity and mortality among such patients. The response of the government and the pharmaceutical industry was immediate and provided financial, technological, and human resources to the field. Evaluations of antifungal therapies and the development of improved and rapid diagnostic tests were then initiated. In conjunction with the development of antifungal agents, the high prevalence of severe and fatal mycoses among these types of patients was ranked as the second most important contribution affecting the development of medical mycology in the United States by the questionnaire respondents (15.1% of 86 responses, Table 3, Appendix A).

The discovery of the perfect states of some fungi and the development of a latex agglutination test for the laboratory diagnosis of *C. neoformans* were listed among the most important contributions by the questionnaire respondents (Table 3, Appendix A). Other basic research contributions of this decade included: achieving a better understanding of the mechanisms of the mycelium-to-yeast conversion by *H. capsulatum* var. *capsulatum* and *S. schenckii*; understanding cellular immunity as a major factor in chronic histoplasmosis capsulati; discovering the role of the capsule of *C. neoformans* in phagocytosis as a virulence factor; and defining the role of neutrophils as effective opportunistic fungal pathogens killers. In applied research, additional adaptations of two bacteriologic procedures, Örjan Ouchterlony's immunodiffusion test and the direct fluorescent antibody procedure contributed new dimensions to the rapid diagnosis by clinical laboratories. Another antifungal agent effective against yeast infections, 5-fluorocytosine, also was developed and evaluated.

By 1955 tuberculosis had began to be managed effectively, and the Veterans Administration and Armed Forces increased their support for the study of fungal diseases. In the early 1960s, multicenter investigations studied large numbers of patients. A leader of this research was John Busey, who initiated the first cooperative drug study that evaluated the effectiveness of amphotericin B and stilbamidine while at the VA Medical Center in Jackson, Mississippi. Also during this period, Elizabeth Hazen retired from the New York State Laboratory in 1960 and was succeeded by Morris Gordon, who had moved there from the CDC.

On November 6, 1961, Lorraine Friedman and Donald Schneidau from Tulane University in New Orleans, Louisiana wrote to John E. Blair, President of the ASM, suggesting the creation of a Medical Mycology Division within ASM. The purpose of this division was to arrange a program centered on medical mycology during the annual ASM meetings and provide a focal point for the activities of medical mycologists (Howard, 1974). In 1969, an *ad hoc* committee was appointed to review the organization of ASM, especially the part governing divisions. In 1972, microbiological groups that had gained the votes of at least 150 members were given the status of ASM divisions. The Medical Mycology Division was then established by the votes of 181 members (American Society for Microbiology Archives, 1973).

The history of medical mycology at the Medical College of Virginia, Virginia Commonwealth University (MCV, VCU) began in 1965, when John P. Utz moved from NIH to MCV, VCU. He brought with him Smith Shadomy and H. Jean Shadomy. These three investigators and Richard J. Duma, who succeeded Utz as Chairman in 1974, established a medical mycology training and research program at MCV, VCU.

In 1965, a steering committee was selected to organize the Medical Mycology Society of the Americas. The members of the committee were: Libero Ajello, Charlotte Campbell, Lorraine Friedman, Milton Huppert, Hillel Levine, and Margarita Silva-Hutner. They met in Phoenix, Arizona during the Second Coccidioidomycosis Symposium and planned an organizational meeting to be held during the ASM Annual Meeting in 1966 (Ajello, 1968). Ajello was the first president.

In 1965, John E. Bennett became Head of the Infectious Diseases Service at the NIH when Utz left (J. Bennett's CV). In Bennett's laboratory, several individuals obtained their postdoctoral training, including two Dalldorf Fellowship trainees ("Training Trees", Appendix B). In 1966, Kyung Joo Kwon-Chung came to Chester Emmons' laboratory at the NIH as a visiting fellow to work on species of the genus *Aspergillus*, as well as on the genetics of the dermatophytes (J. Kwon-Chung, personal communication, March 14, 1994).

Charles E. Smith died on April 18, 1967, and the CDC Kansas City Medical Mycology Unit was closed in 1968. By 1969, medical mycology training and research was being conducted at several other centers: the University of California at San Francisco by Carlyn Halde and Robert Lehrer, the University of Texas at Austin by Paul Szaniszlo, the University of Arizona in Tucson by Jim Sinski, the Harvard School of Public Health by Charlotte Campbell (who had moved from Walter Reed), Columbia University by Irene Weitzman, the University of Chicago by John Rippon, Cornell University by Donald Louria, and Washington University by George Kobayashi and Gerald Medoff. New investigators joined institutions where medical mycology programs had been established: Leo Pine at Duke University, Glenn Bulmer and Harold Muchmore at the University of Oklahoma at Oklahoma City and later at Tulsa, and Gerald Bodey at the National Cancer Institute.

1960 to 1969

Opportunistic fungal infections in immunocompromised patients

Renal transplantation in humans and the use of immunosuppressive therapy to avoid organ rejection began in the early 1960s. The incidence of infectious diseases, including fungal infections, was similar to that observed in patients with various malignancies. David Rifkind, Thomas Marchioro, Stuart Schneck, and Rolla Hill (1967) found at autopsy systemic fungal infections in 23 of 51 patients (45%) who received renal transplants and had immunosuppressive agents between 1962 and 1965 at the University of Colorado Medical Center. Since the beginning of the transplant program in November 1962 at that institution, 111 patients had received renal transplants. Antemortem diagnosis and amphotericin B therapy were possible for only two of these patients.

An increased awareness of the prevalence of mycoses in cancer and leukemia patients was reported in two publications: one from the Memorial Hospital for Cancer in New York City (Hutter, Lieberman, and Collins, 1964) and a second from the National Cancer Institutes (NCI) Leukemia Service (Bodey, 1966). The former authors discovered aspergillosis at autopsy in 30 patients over a 12 year span. Aspergillosis was a contributing factor in the cause of death in 37% of these patients. Gerald Bodey (1966) reviewed the records of 454 patients with acute leukemia, who had died at the NCI between January 1, 1954 and June 3, 1964. He found 189 fungal infections among 161 patients, an important increase in their incidence between 1959 and 1964. These infections were the cause of death in 61% of the patients, who had more prolonged granulocytopenias prior to infection than those patients enrolled in the control group. Both studies emphasized the need for improvement in the laboratory diagnosis of mycoses. Bodey (1966) concluded that histoplasmosis capsulati and cryptococcosis were not as closely related to adrenal cortical steroid therapy as were other fungal infections. This conclusion correlated well with the results from experimentally infected animals (Louria, Fallon, and Browne, 1960) and in vitro studies (Caroline, Rosner, and Kozinn, 1969).

Clinical studies: antifungal therapy

Successful therapy of mycoses with amphotericin B was followed by focused efforts to improve diagnosis, treatment, and the understanding of the different aspects of fungal disease. By 1963, William Winn stated that, "there was the need for systematization of the various names given to coccidioidomycosis, because this infection had many variable manifestations" (p. 1131). Therefore, Winn (1963) developed a system of classification based on radiologic changes in the lungs using his 20 years of experience "derived from almost daily encounters with the pathologic manifestations of coccidioidomycosis" (Winn, 1963, p. 1131). He reviewed data from 100 patients who had been favorably treated with amphotericin B at Tulare-Kings Hospital, California. Winn (1963) divided coccidioidomycosis into three categories: (i) primary (pulmonary, residual pulmonary, extending pulmonary lesions), (ii) acute and chronic disseminated (respiratory to meningeal diseases), and (iii) primary extrapulmonary disease (cutaneous inoculation). The superior value of the intrathecal over the intravenous therapy was emphasized in this publication.

Cooperative studies regarding blastomycosis and other systemic diseases that were initiated by members of the Cooperative Study Group on the Chemotherapy of Tuberculosis of the Veterans Administration-Armed Forces in the 1950s were published in

the 1960s (Espinel-Ingroff, 1994, 1996). Following their successful therapeutic trials for the management of tuberculosis, committees were established to study coccidioidomycosis and histoplasmosis capsulati, and in 1957 to study blastomycosis. The first of these studies (John Busey et al., 1963) synthesized clinical information from a retrospective study of 198 cases of blastomycosis diagnosed during a twelve-year period at 26 Veterans Administration hospitals. Blastomycosis was confirmed to be prevalent in the Mid-Atlantic, South Central, and Ohio-Mississippi River valley states. The study to compare the effectiveness of 2-hydroxystilbamidine and amphotericin B began in 1955 and revealed that amphotericin B was more effective than the former drug (Lockwood, Allison, Batson, and Busey, 1969) in managing this mycotic infection. Eighty patients with blastomycosis were evaluated during a 11 year period; 2-hydroxystilbamidine was used successfully in 31 of 53 patients (58%) and amphotericin B in 21 of 27 patients (78%).

Michael Furcolow (1963) and the members of the CDC Cooperative Mycoses Study Group in Kansas City reported the results of their first large cooperative study on long-term follow-up evaluation of amphotericin B therapy for histoplasmosis capsulati. Comparison of data from 194 treated and 115 untreated patients indicated that the total dose should be at least 25 mg/kg (23% mortality). Inadequately treated patients (<25 mg/kg) showed little difference from untreated patients (85% mortality).

Despite the introduction of amphotericin B and its successful use for the treatment of fungal infections, its toxicity limited its use. David Drutz, Anderson Spickard, David Rogers, and Glenn Koenig (1968) investigated a new therapeutic approach to reduce the dose-related toxicity of amphotericin B at the time when Drutz was doing his postdoctoral training at the NIH. The drug was administered daily for a ten-week period to achieve peak serum levels that were at least twice those necessary for inhibition of the fungus. These lower total doses were sufficient to control 15 systemic cryptococcosis, blastomycosis, and histoplasmosis capsulati infections in patients with impaired renal function. This antifungal regimen was used in the 1970s by John Bennett and his coworkers for their prospective study on the treatment of cryptococcal meningitis (J. Bennett, personal communication, March 17, 1994).

The new oral agent, 5-fluorocytosine (flucytosine [5-FC]), was developed in the early 1960s by Hoffmann-LaRoche in Switzerland (E. Grunberg et al., 1963). It was found to be effective in the treatment of two patients infected with *C. albicans* and *C. neoformans*, as well as in experimentally infections (E. Grunberg et al., 1963; Tassel and Madoff, 1968). The first patient treated with oral 5-FC had candidal sepsis and received 5-FC for 21 days (2 to 4 gm daily), and the second patient had cryptococcal meningitis, and was treated with 2.25 to 9 gm of 5-FC daily for 50 days (Tassel and Madoff, 1968). John P. Utz and his associates (1968) at MCV,VCU treated 15 patients who had cryptococcal infections with oral 5-FC for 14 to 42 days. They noted improvement in 9 of the 11 patients with meningitis. However, relapse occurred in four of the nine patients. By the 1970s, the development of drug-resistant *C. neoformans* isolates during 5-FC therapy was reported by Edward Block, Anne Jennings, and John Bennett (1973). Nevertheless, the utility of this drug was fully recognized in the late 1970s, when it was used in conjunction with amphotericin B for managing cryptococcal meningitis. Synergism between these two antifungal agents was demonstrated in the 1970s, which provided new alternatives for therapy.

Smith Shadomy (1969), a bacteriologist by training, while he was at MCV, VCU reported the first systematic study of the in vitro susceptibilities of 5-FC and amphotericin B against *C. neoformans*. The procedures that he developed during this study became the basis for preclinical in vitro screening of new antifungal molecules at MCV, VCU and other centers throughout the world. His antifungal susceptibility testing procedures served as useful tools in the clinical laboratory until the early 1990s and his pioneer work helped to establish this area within the discipline. Smith Shadomy screened most of the antifungal agents developed until the late 1980s (T. Kerkering; J. Shadomy; J. Utz, personal communications, January 4, 17, and 7, 1994, respectively).

Laboratory diagnostic methods

Rapid, reliable, and reproducible methods for the diagnosis of fungal infectious diseases were developed in the 1960s. The earlier adaptation of Albert Coon's and Melvin Kaplan's (1950) fluorescent-antibody technique by Warren C. Eveland et al. (1957), Gordon (1958), and Ralph Vogel and Joseph

Padula (1958) demonstrated the potential value of this methodology for fungi. Data from these earlier studies led to further investigations at the CDC using this method for the rapid detection and identification of *S. schenckii* (W. Kaplan and Ivens, 1960), *H. capsulatum* (L. Kaufman and W. Kaplan, 1961), and *B. dermatitidis*, *C. immitis*, *C. neoformans*, *H. capsulatum*, and *S. schenckii*, (W. Kaplan and Kraft, 1969) in formalin-fixed and paraffin embedded tissues. The application of the fluorescent antibody technique for the detection of fungal cells in tissue provided a faster laboratory diagnostic method than culture (W. Kaplan, personal communication, January 6, 1994).

The first demonstration of the "occurrence" of cryptococcal antigen (as a serological reactive substance) in body fluids from a patient infected with *C. neoformans* was reported by James Neill, John Sugg, and D.W. McCauley (1951) at Cornell University. Their discovery led to the development of the cryptococcal latex test by Norman Bloomfield, Morris Gordon, and Dumont Elmendorf (1963) at the New York State Health Department. The latter investigators detected the antigens of *C. neoformans* in serum and CSF in nine cases of cryptococcal meningitis. This was an important breakthrough because previous tests for anticryptococccal antibody were negative for 60% of serum specimens from patients with this infection. A high cryptococcal antigen titer was identified as one of the valuable indicators for predicting the outcome of cryptococcal meningitis (Diamond and Bennett, 1974). Other investigators at the VA Hospital, San Fernando, California attempted to develop easier procedures such as a latex test and other immunodiffusion tests rather than the CF test for *C. immitis* infections (Huppert and Bailey, 1965; Huppert, Peterson, Sun, Chitjian, and Derrevere, 1968). However, the CF test remains the "gold standard" among the serologic tests in coccidioidomycosis. H. I. Lurie and W. J. Still (1969), while at MCV, VCU, demonstrated by electron microscopy that the halo surrounding *S. schenckii* yeast cells in tissue was not a capsule, but an artifact formed by an antigen-antibody complex.

In 1969, a new culture medium called dermatophyte test medium (DTM) was developed by David Taplin, Nardo Zaias, Gerbert Rebell, and Harvey Blank for the simplified diagnosis of dermatophytosis involving soldiers during the Vietnam War when skin lesions were heavily contaminated with soil microorganisms. This medium contained a mould inhibitor (cycloheximide), antibiotics (gentamicin and chlortetracycline HCL), and phenol red pH indicator. Growth of a dermatophyte raised the medium's pH as evidenced by a color change from yellow to red.

Taxonomy and classification

The perfect states of several fungal pathogens were described between 1962 and 1967. Lucille Georg, Libero Ajello, Lorraine Friedman, and Sherry Brinkman (1962), at the CDC and Tulane University, described the dermatophyte *Microsporum vanbreuseghemii* and its perfect state. They proposed that this dermatophyte be classified in the genus *Nannizzia* as *N. grubyia*. In 1967, Ajello and Shu-Lan Cheng demonstrated that the granular form of *Trichophyton mentagrophytes* was heterothallic, based upon mating studies using monoascopore cultures. These authors declared that "the ascospores were characteristic of the genus *Arthoderma*" and named the perfect state of *T. mentagrophytes*, *A. benhamiae*, in honor of Rhoda Benham (Ajello and Cheng, 1967, p. 231). Irene Weitzman (1964a), who was at Columbia University, clarified several taxonomic issues concerning the sexual state of *Microsporum gypseum*. Weitzman (1964a) confirmed Phyllis Stockdale's (1963) conclusion that *M. gypseum* was the imperfect state of "at least two cleistothecium-forming species, *N. incurvata* Stockdale and *Gymnoascus gypseum* Nannizzi emend. Griffin" (Weitzman, 1964a, p. 433). From the latter findings, Weitzman (1964b) also showed that pleomorphism in *M. gypseum* was the result of gene mutation and therefore "the term pleomorphism was to be discarded in favor of the term mutation" (p. 203). These results represented a portion of Irene Weitzman's doctoral dissertation compiled under the guidance of Lindsey Olive at Columbia University ("Training Trees", Appendix B; I. Weitzman, personal communication, August 8, 1994).

While Kyung Joo Kwon-Chung was doing her doctoral training with Kenneth Raper at the University of Wisconsin, Kwon-Chung, D.I. Fennell, and Raper (1964) reported the first heterothallic species of *Aspergillus*, which was isolated from soil samples collected in Costa Rica and New Zealand in 1962. The pattern of its conidial structures and character of its cleistothecia placed this fungus in the *A. nidulans* group. Kwon-Chung stated that her medical mycology career started with this discovery

(K.J. Kwon-Chung, personal communication, March 14, 1994). H. Jean Shadomy had observed in 1964, while she was at the NIH, that *C. neoformans* profusely produced true hyphae (J. Shadomy, personal communication, January 17, 1994). She thought she had a "contaminant" or a mutant, but later (1970) she demonstrated that the hyphae had clamp connections that were another growth form of *C. neoformans*. The finding of clamp connections linked *C. neoformans* with the Basidiomycetes, which led in 1975 to the discovery of the sexual stage of this fungus by Kwon-Chung.

The sexual stage of *B. dermatitidis* was discovered by Eugene McDonough and Ann Lewis (1967) (Department of Biology, Marquette University). Their discovery was made while they were investigating the growth characteristics of paired strains of *B. dermatitidis* (Lewis' dissertation research project). The cleistothecia contained light colored asci and resembled other members of the family Gymnoascaceae. In 1968, McDonough and Lewis established the genus *Ajellomyces* to accommodate the teleomorph state of *B. dermatitidis* in honor of Libero Ajello.

Ecology

The first report of the isolation of *B. dermatitidis* from nature appeared in 1961 (Denton, McDonough, Ajello, and Ausherman). Arthur DiSalvo stated that, prior to this report, the ecological niche occupied by this fungus was only presumed to be in the soil; therefore, this paper should be considered a "milestone" of medical mycology. It required 15 years (1986) to show that *B. dermatitidis* could be cultured from natural soil substrates associated with an outbreak of human disease. (A. DiSalvo, personal communication, May 9, 1994). DiSalvo was J. Fred Denton's student at the Medical College of Georgia in Augusta ("Training Trees", Appendix B).

Dimorphism mechanisms

The transformation from the mycelial form to the yeast form of *H. capsulatum* var. *capsulatum* and *B. dermatitidis* was studied by Dexter Howard at the University of California in Los Angeles. Howard (1961) thought that the conversion process of the dimorphic fungus *S. schenckii* was similar to that observed in these two other fungi. However, *S. schenckii* had two different morphological transformations during its transition from the mycelial to the yeast form: "the formation of budding structures at the tips of the hyphae and the formation of chains of oidia and fragmentation of those chains" (p. 467). The latter transformation was similar to that observed in *H. capsulatum* and *B. dermatitidis*. Howard (1961) nullified the prevalent concept that yeasts were only formed directly from conidial cells. Leo Pine and Robert Webster (1962), who were at Duke University, used the synthetic medium developed earlier by Pine and Carl Peacock (1958) to study the conversion of *H. capsulatum*. Pine and Webster (1962) observed that the formation of yeast cells could occur either by direct conversion of microconidia, by budding of hyphal cells, or by "monilial" chain formation. Everett Beneke, Walter Wilson, and Alvin Rogers (1969) at Michigan State University in East Lansing studied the enzymological basis of conversion in *B. dermatitidis* and showed that this fungus produced acid-and alkaline-phosphatases in the yeast form (37°C), but little or none of these enzymes in the mycelial form.

Host defenses against yeasts

Candida albicans

The capacity of neutrophils to phagocytize *C. albicans* also was investigated in the 1960s. *C. albicans* cells were visualized within neutrophils after their intravenous inoculation into mice. Donald Louria and Robert Brayton (1964) at Cornell University found that the candidal cells were phagocytized and survived within the host cells. The candidal hyphae penetrated viable neutrophils in vitro. More virulent strains had greater capacity to grow out of viable leukocytes or circumvent host cellular defenses. These results led to the studies conducted at the University of California, San Francisco by Robert Lehrer and Martin Cline (1969), who measured quantitatively the candidacidal activity of human leukocytes and serum from normal individuals and patients with fungal and other infections. They concluded that the neutrophil played an important role in resistance to *Candida* spp. infections and that the lysosomal enzyme myeloperoxidase and its oxidant substrate, hydrogen peroxide, were the major participants in lethal neutrophil activity. Although serum factors were necessary for optimal phagocytosis, they lacked direct candidacidal activity. It was

confirmed two decades later (1980s) that neutrophils are the most effective killers of most fungal pathogens.

Crytococcus neoformans

Since the 1950s, conflicting results had been obtained regarding the capsule of *C. neoformans* and its significance in phagocytosis. Glenn Bulmer, M.D. Sans, and C.M. Gunn (1967), at the University of Oklahoma, Oklahoma City, established seven non-encapsulated mutants of *C. neoformans* and compared their virulence with that of their encapsulated parents. They observed that the mutant strains were initially avirulent for mice but became virulent in various degrees following reversion to their encapsulated state. Histopathological examination of tissue from human cases of cryptococcosis have shown for many years that giant cells may contain *C. neoformans*. Because the capsule seemed to be a virulence factor, Bulmer and Sans (1967) investigated phagocytosis by using mutants. Phagocytosis, an immunologic defense mechanism, was approximately three times more effective when nonencapsulated mutants were used, which indicated that the polysaccharide capsule inhibited phagocytosis of *C. neoformans*.

Host defenses and Histoplasma capsulatum var. capsulatum

Dexter Howard (1965) studied the effect of environmental conditions upon the intracellular generation time of *H. capsulatum*, which he previously had shown to be constant within mammalian histiocytes. He demonstrated that monocytes from immunized mice did not restrict intracellular proliferation and that hyperimmune serum did not alter this outcome. Howard (1965) concluded that it was "difficult to accept any proposed mechanism of immunity that involved the activity of specific humoral antibodies" (p. 522). In contrast, Marcus Newberry, John Chandler, Tom Chin, and Charles Kirkpatrick (1968) studied the stimulation and transformation by histoplasmin of peripheral blood lymphocytes from both histoplasmin-negative and -positive individuals, as well as patients with either acute or chronic histoplasmosis capsulati (CDC and University of Kansas). Using a more "sensitive indicator" of lymphocyte transformation, the uptake of a radioactive DNA precursor, they observed that lymphocyte transformation was depressed in chronic patients. These results suggested that cellular immunity was a major aspect in chronic histoplasmosis capsulati.

DNA studies

Fungal DNA studies began to be reported in the 1960s. Roger Storck (1966) at the University of Texas, Austin and A. Stenderup and Leth Bak from Denmark (1968) studied the nucleotide composition of DNA from non-pathogenic filamentous fungi and 18 species of the genus *Candida*, including the opportunistic pathogens *C. albicans* and *C. tropicalis*. They found that the mean base composition, most frequently expressed as guanine plus cytosine content, varied among the species studied and that this kind of analysis appeared to have taxonomic and phylogenetic significance.

Physiology and nutrition

John Rippon (1967), while at the University of Chicago, stated that although the mechanisms of disease induction had been investigated for bacteria, the factors involved in fungal infections had been neglected. Rippon conducted important enzymologic studies between 1967 and 1969 and was able to associate the production of elastase with the more virulent forms of dermatophytoses. In 1968, he demonstrated the production of an extracellular collagenase by *Trichophyton schoenleinii*. The severity of ringworm infections was related to the production of enzymes, the latter correlating to mating types (Rippon and Garber, 1969). Rippon believes that these publications were his most important contributions.

According to Rippon, an important contribution to the development of medical mycology was the delineation of the characteristics that separate fungi from other eucaryotic organisms. Fungi biosynthetically synthesize lysine by using the alpha L-aminoadipic acid pathway and have chitin in their cell walls. These characteristics delineate the uniqueness of fungi (J. Rippon, personal communication, March 16, 1994). R.H. Whittaker's (1969) proposal that fungi should be grouped as a separate kingdom was fundamental for the development of medical mycology as a science. The three editions (1974, 1982, 1988) of Rippon's textbook, *Medical Mycology* (Appendix C), also were major contributions,

because the knowledge gained from basic fungal biochemistry, physiology and clinical studies was integrated into a single tome.

Immunotherapy

Morris Gordon and Edward Lapa (1964) found that a globulin fraction of serum from rabbits hyperimmunized to *C. neoformans* significantly enhanced the effect of amphotericin B in experimental cryptococcosis in mice. Arturo Casadevall attempted to adapt this procedure to enhance the treatment of fungal diseases in AIDS patients (A. Casadevall, personal communication, April 10, 1994). This concept aroused a great deal of interest among the members of the NIAID- Mycoses Study Group during their 1994–1996 annual meetings.

Educational and training contributions: 1950 to 1969

Columbia University

1950 was a year of major changes at Columbia University when Gardner Hopkins' retirement ended his years of teamwork with Rhoda Benham, and Lucille K. Georg accepted a position at the CDC. Margarita Silva-Hutner then joined Benham's staff to assume Georg's responsibilities while pursuing her doctorate on weekends with William Weston in the Botany Department at Harvard. Beginning in 1936, Silva-Hutner had worked as a mycology assistant in Arturo L. Carrión's laboratory in Puerto Rico for 11 years, where Carrión encouraged her to continue her studies of chromoblastomycosis under the guidance of his friend, William Weston (M. Silva-Hutner, personal communication, November 20, 1994).

In 1956, newly married Silva-Hutner was appointed Benham's successor and directed the laboratory until her retirement in 1981 (M. Silva-Hutner, personal communication, November 20, 1993). At Columbia, she refined and updated Benham's course by spending its first three weeks teaching general mycology so that the students could become familiar with fungal structures, and then during the last nine weeks concentrating on medical mycology. She included the phaeoid fungi, because they had been the focus of her doctoral dissertation and of her training with Carrión. This course is no longer offered at Columbia (Silva-Hutner, personal communication, 1994; I. Weitzman, personal communication, August 10, 1994).

Michigan State University

An important training program of this era was initiated by Everett S. Beneke at Michigan State University in 1951. A general mycologist by training at the University of Illinois, Beneke took Norman Conant's medical mycology course in 1949 at Duke University (E. Beneke, personal communication, February 22, 1994). In 1951, Beneke began an annual ten-week course in medical mycology for microbiology and pathology graduate students, medical technicians, and physicians. He developed an extensive slide culture collection for the identification of the known medically important fungi. A second year course also was given that concentrated on experimental infections, including extensive tissue work and study of the deep mycoses. During his 40-year career (35 years at Michigan State University), and in workshops offered in various locations in the United States, Beneke trained over 2500 individuals (E. Beneke, Questionnaire). Alvin L. Rogers, who trained with Beneke at Michigan State, teamed with him in offering the courses and training workshops after 1965.

Beneke and Rogers are well known to students of medical mycology as the authors of two important books, *Medical Mycology,* 8th. ed., 1984, the Upjohn Co., Kalamazoo, Michigan and *Medical Mycology Manual* and *Human Mycoses,* 8th. ed., 1996, Star Publishers Co., Belmont, California (Appendix C). These books have been used as standard textbooks for courses dealing with medical mycology including courses for medical students and as reference guides for clinical laboratories since 1968; they served as important tools for furthering the discipline.

New York University Medical Center (NYU)

In 1952, Stanley Rosenthal became Assistant Professor in the Dermatology Department at NYU. Then in 1985, he became the Director of the Mycology Laboratories at the University Hospital and Bellevue Hospital until he retired. He took his first course in medical mycology in 1953 at Columbia University under the guidance of Rhoda Benham and Margarita Silva-Hutner.

The National Institute of Health (NIH) – NCI

The Clinical Cancer Center (now known as the National Cancer Institute [NCI]) was opened at the NIH in July 1953. Gerald Bodey, Robert C. Young, Philip Pizzo, Thomas Walsh, and Emmanuel Roilides are among the scientists who have contributed substantially to the development of medical mycology at this center. In the same year, John P. Utz became Chief of the NIH Infectious Diseases Service and conducted postdoctoral clinical training for physicians (Clinical Associates), while Emmons continued to provide postdoctoral training to the Research Associates. In the Medical Mycology Section at the NIH, Herbert Hasenclever succeeded Chester W. Emmons as Head Mycologist from 1966 to 1974. He had served under Emmons as a Senior Scientist from 1957 to 1965. Hasenclever was a bacteriologist by training and learned his medical mycology in association with Emmons. Hasenclever ably continued the highly regarded and productive work of his mentor, and trained Errol Reiss, who is now at the CDC in Atlanta. Hasenclever died on September 21, 1978, at the age of 54 (Espinel-Ingroff, 1994).

Emmons continued his medical mycology contributions during the 1950s and 1960s. The result of such a wealth of experience and knowledge was his leading textbook, *Medical Mycology*, published in 1963 with Chapman Binford and John P. Utz at the invitation of Lea and Febiger (Appendix C). Emmons edited three editions. This classic grew out of Friday afternoon "Learning and Teaching" sessions with Binford and others at the Armed Forces Institute of Pathology (J. Utz, personal communication, January 7, 1994; Campbell, 1983). Utz was asked to provide needed clinical insight, especially in the diagnostic and therapeutic areas. Numerous other published reports and findings originated from the Friday afternoon discussions, many of them breakthroughs in the United States and abroad. From 1954 to 1965, Utz trained the clinical associates via morning grand rounds and afternoon laboratory sessions. Up to 18 clinical associates were at the NIH at any one time, but only a few were interested in medical mycology (J. Utz, personal communication, January 7, 1994). In 1963, H. Jean Shadomy, who had trained and worked with Plunkett at UCLA, joined Utz to direct his laboratory until they moved to the MCV in 1965.

In 1961, the physician John Bennett came to the NIH as a Clinical Associate after finishing his house-staff training. He was given permission to work and train with Emmons for three years and he also substituted for Utz during the latter's sabbatical year at the Institut Pasteur in Paris, France. In 1965, Bennett took over Utz's position and in 1966, Emmons brought Kyung Joo Kwon-Chung to the NIH as a "Visiting Fellow" even though she was not an American citizen, to help him describing the genus *Aspergillus* for a manual that he was preparing. Kwon-Chung obtained her M.S. and Ph.D. degrees at the University of Wisconsin, Department of Bacteriology under the guidance of the general mycologist Kenneth Raper, a student of John Couch. When Emmons retired in 1968, the mycology laboratory was dismantled and only Kwon-Chung was left in the virology laboratory. Later she moved to Bennett's section beginning then their collaborative work (K.J. Kwon-Chung, personal communication, March 13, 1994).

Tulane University

Morris F. Shaffer (1985) was committed to developing a "strong" medical mycology program at Tulane University, and in 1955 hired Lorraine Friedman, one of Conant's students. Friedman was working as a research mycologist at the Naval Biological Laboratory, University of California, under the supervision of Charles E. Smith. She had obtained her doctoral degree in 1951 under the mentorship of Norman Conant at Duke University ("Training Trees", Appendix B). Friedman soon established a sound training component by teaming with other faculty and departments, especially with Dermatology at both Tulane and the Louisiana State University Medical School. Friedman and Shaffer worked together in 1958 to obtain the first NIH mycology training grant, which was renewed annually for 30 years (Shaffer, 1985; J. Domer, questionnaire). Friedman worked with other faculty members to develop a demanding entry process for students, the result being a number of graduates who have become well known and have made important contributions to the development of medical mycology (Shaffer, 1985; Research section). Among the faculty members, J. Donald Schneidau came to New Orleans in 1952 when he was appointed Lecturer in Mycology in the Department of Tropical Medicine and Public Health. He became a Research Associate

in the Department of Microbiology at Tulane University in 1954. Schneidau earned his doctoral degree in 1956 under the direction of Morris Shaffer and collaborated with Friedman until he retired in 1978. He died on June 19, 1994. Angela Restrepo, a well known mycologist from Medellin, Colombia was Schneidau's student. Judith Domer, another graduate under Morris Shaffer and an associate with Friedman, became Director of the Mycology Program in 1980 when the latter retired. (J. Domer, CV; "Training trees", Appendix F).

Friedman's pre-doctoral program soon attracted a number of students due to its comprehensive curriculum and the opportunity to study with the medical faculty, other leading faculty members, and outstanding scientists who were frequently brought to the University. Each graduate student was required to either develop a dissertation or a related paper worthy of publication by a journal in her or his field. Medical mycology, due to its strong ties to the medical schools and centers for tropical medicine, was an important component of the training program. All pre-doctoral trainees were expected to take Friedman's medical mycology graduate course, a year-long program that encouraged individual research and initiative. They also were required to rotate among the diagnostic, general mycology, dermatology, and clinical laboratories. Friedman stayed in close touch with each student to guide their work and serve as a mentor when needed (Shaffer, 1985).

In addition to diagnostic work, Friedman's laboratory became a training and research nucleus for dermatology residents and graduate students. She led the field in training the first generation of molecular-based medical mycologists ("Training Trees", Appendix B; G. Kobayashi personal communication, July 1, 1993). Included among a long list of her graduates are a number of important medical mycologists: George Kobayashi at Washington University, St. Louis; William Cooper who was at Baylor University, Dallas until his death; Geoffrey Land at the Methodist Medical Center in Dallas; and Thomas G. Mitchell (Conant's successor at Duke University) ("Training trees", Appendix B).

University of California at Los Angeles (UCLA)

Carlyn Halde considers Orda Plunkett's most important contribution while he was at this institution to be his training of medical technologists during the 1950s. He trained about 50 individuals each year (C. Halde, personal communication, December 9, 1993). J. Walter Wilson and Plunkett coauthored the textbook, *Fungous Diseases of Man*, its first edition was published in 1965. The book contains excellent reproductions of Wilson's drawings and photographs from Plunkett's slide collection of fungal structures. It proved to be an outstanding tool (Appendix C).

In 1956, Dexter Howard, a recent bacteriology graduate and new faculty member in the School of Medicine also began teaching medical mycology at UCLA. Howard had studied medical mycology with Plunkett, whom he considers his mentor (D. Howard, questionnaire), but his course work was focused in bacteriology under Gregory Jann. Howard either taught or managed the mycology section of the microbiology course for medical students with help from teaching assistants and other faculty until 1988 when it was discontinued. McVickar had taught this section of the course before Howard. In addition, Howard organized and taught medical mycology courses during the weekends with N. Dabrowa, who was his graduate student. Those courses were given since the late 1970s to medical technologists living in the area. During his years at UCLA until his retirement in 1996, Howard supervised or provided guidance to over sixty doctoral students, postdoctoral fellows, special fellows, and research associates (D. Howard, questionnaire and CV; "Training trees", Appendix B).

University of Oklahoma

George Cozad obtained his masters degree under Michael Furcolow's and Howard Larsh's direction at this institution. He then went to Duke University and carried out his doctoral studies with Norman Conant and received his Ph.D. in 1953. In 1957, Cozad returned to Norman as a Research Associate in medical mycology to work with Larsh, and in 1960, he became Assistant Professor in the Department of Botany and Microbiology. Cozad taught a medical mycology course to premedical, senior undergraduate, and a few graduate students until just before his death (September 29, 1995). Cozad also supervised the graduate studies of several individuals and was the major professor of Juneann Murphy (Ph.D.) and Julie Rhodes (M.S.). Murphy

received her Ph.D. degree in 1969 and is currently at the University of Oklahoma, Oklahoma City, directing a very active basic research laboratory dedicated to studies regarding host defenses against *C. neoformans*. In 1960, Glenn Bulmer became a faculty member at the VA Medical Center in Oklahoma. He had obtained his training with Everett Beneke at Michigan State University and in 1965, he obtained a career development NIH grant to study phagocytosis in *C. neoformans* infections and the role of the capsule of this organism as a virulence factor. Robert B. Fromtling, who did his postdoctoral studies under the Shadomys at MCV,VCU, was one of Bulmer's students ("Training Trees", Appendix B). During this decade, Harold Muchmore also came to the University of Oklahoma as a faculty member in the Medical Microbiology Department, where he taught medical mycology to graduate and medical students, provided medical mycology training, and conducted important research studies (See Scientific Contributions). Both Bulmer and Muchmore had retired and Larsh died in 1994. None of these individuals have been replaced. Individuals who trained under the above medical mycologists at the University of Oklahoma are listed in the "Training Trees", Appendix B.

Temple University

In 1962, Fritz Blank moved to the Skin and Cancer Hospital at Temple University School of Medicine from McGill University in Montreal. He established a medical mycology training and research program with the support of Brown-Hazen grants that he had obtained since 1958; he also maintained a diagnostic mycology laboratory. Blank was a German-born chemist whose early training in Basel, Switzerland with Hans Scholer on the chemotherapy of fungal infections had brought him into the field. In addition, Blank had studied medical mycology in the laboratory of the famous French mycologist, Raimond Sabouraud, at the Hospital St. Louis and the Institute of Parasitology in Paris, France.

Medical College of Virginia, Virginia Commonwealth University (MCV, VCU)

In 1965, Utz, J. Shadomy and her husband Smith Shadomy moved to MCV, VCU in Richmond, Virginia to establish the Division of Infectious Diseases in the Department of Medicine. Medical mycology was an important element of their training and research for the next 28 years. Much of their initial work was supported by NIH grants, as well as Brown–Hazen training grants in the 1970s (Baldwin, 1981). In 1968, J. Shadomy began offering a graduate course in medical mycology for dermatology residents, microbiology students interested in medical mycology, postdoctoral students, fellows in infectious diseases, and clinical laboratory personnel; S. Shadomy, a self-taught medical mycologist, was the other faculty member. This three-credit course was given in the Department of Microbiology every other year in conjunction with courses in bacteriology, virology, and parasitology. It covered the clinical and diagnostic laboratory aspects of the pathogenic fungi which were provided during one hour lectures and two hour laboratory sessions. In addition, the Shadomys and Utz taught medical mycology, which was introduced into the MCV, VCU curriculum, to the medical and dental school students.

Centers for Disease Control and Prevention (CDC)

Between 1950 and 1951, Lucille Georg and William Kaplan joined Libero Ajello at the CDC. Kaplan had trained as a veterinarian at Cornell University, but he received his medical mycology training with Ajello and Georg. Kaplan participated as a faculty member, until he retired in the 1980s, in the various medical mycology courses and workshops that Ajello and his group offered at the CDC. By 1959, Leo Kaufman had moved from the University of Kentucky to began his career at the CDC, where he established the National Center for Fungal Serology in 1964. Kaufman also obtained an appointment at the Georgia State University, Department of Microbiology and at the Emory University, Department of Dermatology. In addition, he directed the studies of ten doctoral candidates, including Paul Standard. In 1963, Leo Pine joined Ajello's team after working in the laboratories of Chester Emmons at the NIH, Norman Conant at Duke University, and in 1963 with Edouard Drouhet, Francois Mariat, and Gabriel Segretain at the Institut Pasteur in Paris, France. Pine was awarded MMSA's Benham Medallion in 1987 and died on December 29, 1994. The names of the individuals who trained under these investigators are listed under Appendix B ("Training Trees", Appendix B) and their research contributions under the scientific sections. The first CDC *Laboratory*

Manual for Medical Mycology was published in 1963 (Appendix E). Co-authored by Libero Ajello, Lucille Georg, William Kaplan, and Leo Kaufman, this manual instantly became a major training tool and reference guide. Most state laboratories and many hospitals and university centers used the manual extensively as a self-teaching instrument and reference "bible".

Beginning in 1968, Leanor Haley, who had been Norman Conant's first doctoral student, came to the CDC as Chief of the Clinical Medical Mycology Training Branch. Hilliard Hardin preceded Haley in this position as the demand for more medical mycology training led to the creation of this training program. Haley's charge was to bring training opportunities to health centers and hospitals across the United States where they were needed. She was asked to focus on the clinical aspects of fungal diseases and she modified the existing CDC training program to include histopathology and differential diagnosis. Haley initiated three short courses and workshops (e.g., three days) on specific topics such as serology, etc. The four-week course covered all the pathogenic fungi and included lectures, identification of unknown cultures and slides, and studies of clinical problems and case reports (last week). The two week course on the dermatophytes covered the isolation and identification of these fungi in the clinical laboratory as well as their clinical lesions and therapy. An advanced clinical medical mycology course included clinical cases, histopathology, and the identification and study of rare systemic and opportunistic fungal pathogens. Students were divided into teams that worked on a case history to present in class for review (L. Haley, personal, communication, December 31, 1993).

An important part of Haley's training curriculum was her handbook, *Laboratory Methods in Medical Mycology* (Appendix C). It was first published in 1964 under the title, *Diagnostic Medical Mycology*, while she was at Yale University. This manual enabled laboratory personnel to duplicate in the laboratory what Norman Conant and his coauthors stated was necessary to diagnose fungal diseases rapidly and accurately, as well as the fungi causing them (L. Haley, personal communication, December 31, 1993). Like the other CDC faculty, Haley specialized in training public health graduate students who conducted projects at the CDC (approximately 30, including 15 postdoctoral students). Individual bench training, tailored to the needs of the students, was given that included unknowns, case histories, histopathologic material, and daily one-on-one sessions. In the meantime, Ajello and his team continued teaching short, three day, specialized training workshops.

University of Arizona

James T. Sinski came to this university in 1966 as Associate Professor in the Department of Microbiology and Immunology. He obtained his Ph.D. degree under the direction of the general mycologist, John S. Karling, at Purdue University. Sinski carried out his postdoctoral medical mycology training at Tulane University at Lorraine Friedman's laboratory. During Sinski's tenure at the University of Arizona (1966 to 1996), he taught general and medical mycology graduate courses and received the Outstanding Teacher Award for several years. The individuals who trained under Sinski are listed under Appendix B ("Training Trees", Appendix B). In the 1970s, the physician John Galgiani became a faculty member at the VA Hospital and established a medical mycology research laboratory at this institution. Galgiani conducted his postdoctoral medical mycology training at the laboratory of David Stevens (See Chapter VI).

V. The years of expansion: 1970 to 1979

Figure 81. Herbert F. Hasenclever (pp. 89, 96) and Kyung Joo Kwon-Chung (pp. 89-90) in 1970 (Courtesy of Dr. Sinski)

Figure 82. Errol Reiss (pp. 89, 96-97)

Figure 83. Michael McGinnis (3rd) (pp. 89, 90, 98), his mentor Libero Ajello, and Arvind Padhye at CDC (pp. 87, 97) (Courtesy of Dr. McGinnis)

Figure 84. Hillel Levine (p. 91) (Courtesy of Dr. DiSalvo)

Figure 85. Norman Goodman's Mycology Workshop in 1972. Seated: Glenn Roberts (5th); Third row: Ernest Chick (1st), Howard Larsh (5th), and Norman Goodman (6th) (pp. 94-95) (Courtesy of Dr. Goodman)

Figure 86. Faculty Mycology Workshop in 1970s. Seated: Norman Goodman (2nd) and Norman Conant (3rd). Second row: Ernest Chick (1st), and Michael Furcolow (2nd) (pp. 94-95) (Courtesy of Dr. Goodman)

Figure 87. Fritz Blank (first row, 3rd). First row: Jan Schwarz (1910-994) (1st) and Ernest Chick (4th); Second row: Morris Gordon (1st) and Sarah Grappel (3rd) (p. 95) (Courtesy of Dr. Gordon)

Figure 88. Helen Buckley and John Rippon (p. 95) (MMSA Archives)

Figure 89. Eric Jacobson and his student Karin Nyhus (p. 96) (Courtesy of Dr. Jacobson)

Figure 90. Gilbert Dalldorf (3rd) and first Dalldorf Fellow, Thomas Kerkering (2nd) and his mentors Smith Shadomy (1st) and H.J. Shadomy (p. 96) (Courtesy of Dr. H.J. Shadomy)

Figure 91. Richard Calderone (1st) (p. 97), Nina Dabrowa, and Dexter Howard (MMSA archives)

Figure 92. Milton Huppert (p. 98) (Courtesy of Dr. Rippon)

Figure 93. Bill Cooper (pp. 98-99) (Courtesy of Dr. Domer)

Figure 94. Elmer Brummer (p. 100)

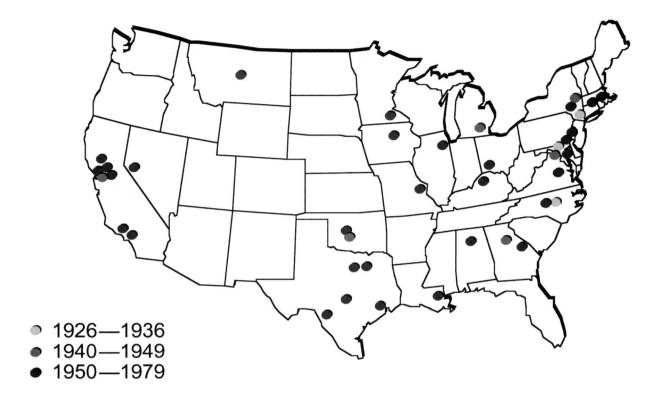

Figure 95. Map of the United States showing the training and research centers up to 1979 (Designed by the author).

Overview: 1970 to 1979

During the years of expansion, the science of medical mycology flourished in research and training as major scientific events changed and enlarged the importance of fungal diseases in clinical medicine. There was an increase in the spectrum of available antibiotics, immunosuppressive and cytotoxic therapeutic agents, and their use. The latter two targeted the hematopoietic cells of the neutrophil lineage in the host and rendered patients highly susceptible to opportunistic fungal infections. The increased incidence and prevalence of fungal infections led to the development of numerous rapid tests for the detection of fungi in clinical specimens and the identification of fungi in the clinical laboratory. Techniques for the direct demonstration of *Candida* spp. cells in blood samples were improved and commercial yeast identification systems for the differentiation of medically important yeasts recovered in the clinical laboratory were developed, e.g., the API 20C Yeast Identification System. Ira Salkin stated that the API 20C was the first commercial product that permitted rapid and accurate identification of the majority of pathogenic yeasts (I. Salkin, questionnaire).

When it became evident that the detection of antibodies was not a reliable laboratory tool for the diagnosis of invasive aspergillosis and candidiasis, alternative approaches began to be investigated like the detection of fungal antigens and metabolites. The identification of dimorphic fungi in vitro was simplified by the development of the exoantigen test. The availability of this test and other simplified procedures stimulated interest in medical mycology among microbiologists and physicians, and demonstrated the need to develop training programs, workshops, and courses. These tests simplified diagnosis, improved patient care, standardized methodology for defining fungi, and made yeast identification cost effective in the laboratory. Rapid and commercially available tests (serologic and non-serologic) were considered a major contribution to the development of the discipline by 24.7% of the questionnaire respondents (Table 4, Appendix A).

Data from large, cooperative therapeutic trials established the regimen of choice in treating cryptococcal meningitis. Ketoconazole, an important azole antifungal agent, was developed by Janssen Pharmaceuticals in Belgium. In the late 1970s, the Mycoses Study Group, later known as the National Institute of Allergy and Infectious Diseases Mycoses Study Group (NIAID-MSG), was formed under the sponsorship of NIAID. These MSG physicians saw the need to conduct systematic studies of fungal diseases and their therapies by developing collaborative protocols. John Bennett, Richard Duma, Gerald Medoff, and William Dismukes met informally at Chicago's O'Hare Airport in the late 1970s and selected Dismukes from the University of Alabama in Birmingham as the chair of the group. The first NIH grant for these studies was awarded in 1978 (W. Dismukes, personal communication, January 8, 1994; M. Saag, personal communication, December 21, 1993). Michael Saag, who trained under Dismukes ("Training Trees", Appendix B), stated that the effect of the NIAID-MSG contributions to the development of the discipline was dual, medical and global. From the "medical point of view, its main contribution was the evaluation of oral antifungal agents for the treatment of systemic mycoses". From the "global point of view, NIAID-MSG crystallized thinking about the importance of systemic mycoses and brought attention to the field" (M. Saag, personal communication, December 21, 1993). The NIAID-MSG heightened awareness of the discipline and focused attention upon fungal diseases. It has served as a catalyst and standard for conducting multicenter, prospective clinical trials, and has emphasized the advantages of collaborative clinical investigations. Since then, the management and understanding of fungal diseases have been enhanced by this group of physicians (W. Dismukes, personal communication, January 15, 1994; J. Perfect, questionnaire).

Pathogenesis and host-parasite interaction studies continued in the 1970s with the development of animal models to investigate the mechanisms of phagocytosis and cell-mediated host defense. Analysis of the immunochemistry of both mycelial and yeast forms of *B. dermatitidis*, *C. albicans*, *H. capsulatum* var. *capsulatum* and the major antigens of *P. boydii* were performed. The perfect states of other fungal pathogens were described, and the clinical nomenclature and taxonomy of human pathogenic species of phaeoid fungi were clarified. The discovery of the sexual states permitted further genetic studies and more accurate taxonomic treatments reflecting phylogenetic relationships.

From 1972 to 1990, the Brown-Hazen Fund was devoted exclusively to support research and training in medical mycology. Among the several scholarships granted, the Dalldorf Postdoctoral Fellowship was awarded annually (1978–1988). The purpose of the Fellowship was to encourage individuals with appropriate education and training to undertake research careers in medical mycology by obtaining specialized training and experience (Research Corporation, 1990).

A postdoctoral research and training program in medical mycology was established in 1972 at Washington University. In the fall of 1974, Thomas Mitchell succeeded Norman Conant as Director of the Medical Mycology Laboratory at Duke University (T. Mitchell, personal communication, January 12, 1994). Wiley Schell, who trained under Michael McGinnis at the University of North Carolina, was recruited to supervise the Clinical Mycology Laboratory at Duke University. In 1977, David Durack came to Duke as Chief of the Division of Infectious Diseases and recruited John Perfect. Perfect, a clinician-researcher, has focused on experimental cryptococcosis and molecular studies, and as an active member of the NIAID-MSG, he has participated in several clinical trials conducted by this group.

By the end of 1979, medical mycology research and training was conducted at a number of other centers: University of Kentucky by Ward Bullock and Norman Goodman, Santa Clara Valley Medical Center by David Stevens, University of Arizona by John Galgiani, University of Alabama by William Dismukes, University of Texas MD Anderson Cancer Center by Roy Hopfer and Gerald Bodey, University of North Carolina at Chapel Hill by Michael McGinnis, University of Texas Health Science Center at San Antonio by David Drutz, Milton Huppert, Marc Weiner, and John Graybill, University of California, Davis by Paul Hoeprich and Demosthenes Pappagianis, Johns Hopkins University by William Merz, and the Mayo Clinic by Glenn Roberts. New investigators who became active in the field at recognized centers included: Juneann Murphy (University of Oklahoma-Norman), Errol Reiss and Arvind Padhye (CDC), Carol Kauffman (VA Medical Center – University of Michigan-Ann Arbor), Ana Espinel-Ingroff, Eric Jacobson, and Thomas Kerkering (Medical College of Virginia), and Judith Domer (Tulane University). Elizabeth Hazen, one of the two discoverers of nystatin, died on June 24, 1975.

1970 to 1979

Immunocompromised patients versus opportunistic fungal infections

Review of all autopsies (3374) conducted at the NIH from July 1953, when the Clinical Cancer Center opened, until January 1968 disclosed 98 cases of fatal aspergillosis (Young, Bennett, Vogel, Carbone, and De Vita, 1970). The most perplexing feature of these cases of aspergillosis was the difficulty in culturing the fungus antemortem (only 34%). Richard Meyer, Lowell Young, Donald Armstrong, and Bessie Yu (1973) found that cases of aspergillosis doubled between 1969 and 1970 at the Memorial Sloan-Kettering Cancer Center in New York City

Carol Kauffman et al. (1978) reported 58 histoplasmosis capsulati cases in immunosuppressed patients between September 1953 and December 1957 at the Indiana University Medical Center and the University of Cincinnati Medical Center. Serological diagnosis was not helpful among these patients.

A higher incidence of disseminated *C. tropicalis* infections, which previously had been associated with only 20 to 25% of candidiasis cases, began to appear in the literature. John Wingard, William Merz, and Rein Saral (1979) found that 15 of 18 candidiasis cases were caused by *C. tropicalis* during a 12-month period at the Johns Hopkins Oncology Center. Merz, who did his postdoctoral work at Columbia University with Margarita Silva-Hutner, stated that the creation in the late 1970s of the Oncology Unit at Johns Hopkins stimulated medical mycology research at that institution. Merz emphasized the importance of working in collaboration with physicians in order to produce laboratory and clinical data (W. Merz, personal communication, January 5, 1994). This collaborative group demonstrated for the first time that surveillance studies are important and that colonization with *C. tropicalis* results in problems among oncology patients.

Phaeohyphomycosis

The new species *Phialophora parasitica* was isolated from a subcutaneous infection in a kidney transplant patient by Libero Ajello, Lucille Georg, Roy Steigbigel, and C.J.K. Wang in 1974. The kidney transplant patient, who was on immunosuppressive maintenance therapy at Stanford University Hospital, developed an abscess on the left arm, which drained small amounts of a yellow exudate (Ajello et al., 1974). The lesion increased in size, and the entire granulomatous mass (1 cm) was surgically removed, after which, the patient was free of further infection. Biopsy material revealed numerous granulomata in the dermis, with a central core of necrosis that contained septate, branched, phaeoid hyphae. Fungal colonies were white at the periphery with a grayish brown center. Microscopic examination revealed "the presence of numerous hyaline, ovoid to cylindroid conidia and simple phialides without collarettes" (Ajello et al., 1974, p. 493). The fungus was named *P. parasitica* because it resembled other members of the genus *Phialophora*. The term phaeohyphomycosis was proposed by Ajello et al. (1974) as a collective name for mycoses caused by several species of phaeoid fungi, when phaeoid hyphal tissue-form cells were seen in contrast to the muriform tissue cells of chromoblastomycosis. A clear distinction was made between chromoblastomycosis, which is also caused by phaeoid fungi, and the subcutaneous abscesses seen in their patient. Historically, these two types of infections would have been described under the single name chromomycosis; but the latter term has been rejected by most mycologists.

Epidemiology

The prevalence and higher incidence of fungal infections continued being documented during the 1970s. Michael Furcolow, Ernest Chick, John Busey, and Robert Menges (1970) at the University of Kentucky reviewed all proven cases of blastomycosis in humans (1476 cases) and dogs (384 cases) in the United States and confirmed earlier reports by the Veterans Administration – Armed Forces cooperative study, that had delineated the endemic areas of this disease.

Clinical studies: Antifungal therapy

In the beginning of the 1970s, two antifungal drugs were available for the systemic treatment of fungal diseases, amphotericin B and 5-FC. Amphotericin B had toxic side effects, and 5-FC was effective only against certain yeasts. Antibacterial therapy had been improved through the synergistic effect of two antibiotics. Antifungal therapy was also improved when at Washington University, Gerald Medoff, M. Comfort, and George Kobayashi (1971) discovered

that the synergistic antifungal activity of amphotericin B and 5-FC against yeasts in vitro was ten times greater than the activity of each drug alone. The synergistic mechanism was the result of the increased entry of 5-FC into yeast cells that was potentiated by amphotericin B (Medoff, Kobayashi, Kwan, Schlessinger, and Venkov, 1972). The synergistic effect of amphotericin B and 5-FC also was demonstrated in experimentally infected mice by E. Titsworth and E. Grunberg (1973), who worked at Hoffman-LaRoche Nutley in New Jersey.

Based on the reports originating from Washington University regarding the synergistic activity of 5-FC and amphotericin B, two collaborative studies were designed to evaluate this combination in the treatment of human cryptococcosis. The first study was an uncontrolled prospective evaluation conducted from May 1971 to November 1973 at MCV, VCU in Richmond and the University of Virginia in Charlottesville by John P. Utz et al. (1975). Fifteen of the 20 patients treated had *C. neoformans* meningitis; of these 15 patients, 12 were cured and no patients relapsed. These encouraging results led to a second prospective, randomized collaborative study involving 10 institutions to compare six weeks of combined amphotericin B and 5-FC therapy with six weeks of amphotericin B therapy alone in 50 patients with cryptococcal meningitis (Bennett et al., 1979). The amphotericin B regimen was similar to that advocated by David Drutz et al. in 1968. Because the combined therapy produced fewer failures and relapses (3 versus 11), sterilized the CSF more rapidly, and was less nephrotoxic than amphotericin B alone, John Bennett et al. (1979) concluded that, "this is the regimen of choice for cryptococcal meningitis" (p. 131). However, during the 1980s, *C. neoformans* infections in AIDS patients were less responsive to this regimen, which stimulated the search for alternative therapeutic regimens for this and other fungal infections in this patient population.

Fungal molecular genetics

The first report of the successful release of living protoplasts from albino and brown types of *H. capsulatum* var. *capsulatum* was published by Martha Berliner and Maria E. Reca (1970), who were at the Harvard School of Public Health. The production of spheroplasts was important for studying the cell wall composition, physiology, cytology, and genetics of fungi. At the University of Houston, Ayodele Olaiya and Stephen Sogin (1979) compared microfluorometrically the flow of the DNA content of haploid, diploid, triploid, and tetraploid strains of *Saccharomyces cerevisiae* with the DNA content of *C. albicans*. They found that *C. albicans* contained a diploid amount of DNA. This discovery ruled out the preconceived idea that *C. albicans* was a haploid yeast like the other species of this genus and led to important investigations on the genetics of fungi during the 1980s and 1990s.

Immunochemistry

The immunochemistry of fungal products also was studied in the 1970s by Judith Domer and Errol Reiss, among others. Domer obtained her Ph.D. degree at Tulane University where her special interest in histoplasmosis capsulati was intensified when she acquired this disease in the laboratory. She recovered under the medical care of John Seabury. Domer (1971) investigated the content of the mycelial and yeast forms of *B. dermatitidis* and *H. capsulatum* var. *capsulatum* and found two chemotypes of this variety based on chitin and glucose content. She found no evidence of chemotypes in *B. dermatitidis*. R.E. Syverson, Helen Buckley, and Charlotte Campbell (1975) at Harvard School of Public Health, on the other hand, compared the cytoplasmic antigenic composition of the two morphologic forms of *C. albicans* by two-dimensional crossed line electrophoresis. They found six distinct soluble antigenic constituents in each form that were not present in the other form.

Errol Reiss, S.H. Stone, and Herbert Hasenclever (1974) demonstrated that the cell walls of *C. albicans* contained three cell wall fractions and mannan, while Reiss was conducting his postdoctoral research in Hasenclever's Laboratory at the NIH. The cell wall fractions, peptidoglucomannan, soluble mannoglucan, insoluble mannoglucan, and mannan, were antigenic and elicited delayed-type hypersensitivity, as measured by skin test and specific inhibition of macrophage migration. The structure of the cell wall mannan separated *C. albicans* into two serotypes, A and B (Hasenclever and W. Mitchell, 1961). David Lupan and John Cazin (1976) at the Universities of Nevada and Iowa isolated and characterized the two major active antigens from culture filtrates of *P. boydii* (antigen 1) and its anamorph, *Scedosporium apiospermum* (as

M. apiospermum) (antigen 2), by ion-exchange chromatography. Both antigens were immunologically identical by immunodiffusion tests. Lupan and Thomas Kozel were Cazin's students at the University of Iowa ("Training Trees", Appendix B). The pathogenicity of this fungus was confirmed in both the sexual and asexual forms (Lupan and Cazin, 1973).

Taxonomy and classification

The perfect states of three fungal pathogens were discovered between 1972 and 1975 by Kyung Joo Kwon-Chung at the NIH. Chester Emmons wanted Kwon-Chung to study the genetics of the dermatophytes, because medical mycologists were not trained in the genetics of other human pathogens. However, she focused on the more invasive fungal pathogens. By 1969, Kwon-Chung had studied the cytology of *C. immitis* and discovered that it had three chromosomes (J. Kwon-Chung, personal communication, March 14, 1994). She then studied the ecology of *H. capsulatum* var. *capsulatum* and discovered the sexual reproductive cycle of this fungus in 1972. Kwon-Chung (1972) identified the two mating types of *H. capsulatum* (+ and –) among natural as well as clinical isolates and named its teleomorph, *Emmonsiella capsulata*, in honor of Chester Emmons. As a sequelae of this discovery, Kwon-Chung (1975a) reported that the "sexual state of *H. duboisii* was identical to *Emmonsiella capsulata*" (p. 980), because cleistothecia and ascospores produced by both mating types of *H. capsulatum* and *H. duboisii* were identical to those of *E. capsulata*. Therefore, she treated *H. duboisii* as a variety of *H. capsulatum* rather than a separate species. In 1979, Michael McGinnis and Barry Katz transferred *E. capsulata* to the genus *Ajellomyces*, which was named to honor Libero Ajello.

From observations reported by Jean Shadomy (1970) at MCV, VCU and Kyung Joo Kwon-Chung's knowledge that many fungi have a sexual state, the latter found a basidiomycetous form in four pairs of *C. neoformans* isolates after a 3-week incubation period: 1 week at 25°C and 2 weeks at 15°C (Kwon-Chung, 1975b). In her search for the sexual stage of *C. neoformans*, J. Shadomy had used only strains isolated from humans. Kwon-Chung, on the other hand, thought that perhaps these isolates belonged to only one of the mating types. The latter investigator decided to test environmental strains and used John Bennett's collection of environmental isolates from Denmark. In order to have a stable *C. neoformans* sexual state, Kwon-Chung developed a sucrose (40%) agar medium that contained all the ingredients useful for fungal growth. The basidial structures found by Kwon Chung (1975b) showed a close relationship with the genus *Filobasidum*, a basidiomycetous genus described by Lindsay Olive. She observed enough differences to warrant the description of a new genus. Kwon-Chung "knew that she had a new genus because it did not look like anything else that she had seen before" (J. Kwon-Chung, personal communication, March 14, 1994). She then discussed her discovery with Olive, who had transferred from Columbia University to the University of North Carolina at Chapel Hill, and established the new genus *Filobasidiella* in 1975 based upon *F. neoformans*. It is now known that *F. neoformans* has two varieties, var. *neoformans* (anamorph, *C. neoformans* var. *neoformans*) and var. *bacillisporus* (anamorph, *C. neoformans* var. *gattii*) (Kwon-Chung, 1992; personal communication, March 14, 1994).

In 1977, the fungal taxonomist Michael McGinnis, while he was at the University of North Carolina, Chapel Hill, summarized and established the clinical nomenclature and clarified the taxonomy of nine human pathogenic species of the genera *Exophiala*, *Phialophora*, and *Wangiella*. Saccardo's original classification "was based upon an artificial system that does not consider natural relationships" (McGinnis, 1977, p. 37). The traditional systems had created a great deal of confusion due to the pleomorphic nature of many of the phaeoid fungi and the variation in generic and specific concepts used by earlier taxonomists. McGinnis provided excellent descriptions of the taxonomic features of each species, which were accompanied by photomicrographs useful for the identification of these fungi in the clinical laboratory.

Laboratory diagnostic methods

The high incidence of *Candida* spp. infections led to the development of improved and efficient tools for the diagnosis of candidemia in the clinical laboratory. These new diagnostic tools were evaluated by medical mycologists working in service laboratories, thereby developing applied research data. Conventional blood culture techniques were compared with the Sterifil lysis-filtration (Millipore Corp., Bedford, MA) by Richard Komorowski and Silas Farmer (1973); commercially prepared vacuum blood-cul-

ture bottles (Albimi and Difco) were evaluated by Nelson Gantz, Judith Swain, Antone Medeiros, and Thomas O'Brien (1974); biphasic and vacuum blood culture bottles were compared by Glenn Roberts, Carlyle Horstmeier, Marsha Hall, and John Washington (1975); the diagnostic efficiency of two commercially prepared biphasic media (BBL Microbiology Systems) were assessed by L. M. Caplan and William Merz (1978); and the diagnostic value of a new blood culture centrifugation technique was examined by Gordon Dorn, Geoffrey Land, and George Wilson (1979). Data from the evaluation of these new systems indicated that vented biphasic blood culture bottles, venting of broth cultures, and the centrifugation technique yielded more rapid and optimal recovery (83%) than the conventional broth culture technique (67%). During the 1980s, the commercially available lysis centrifugation isolator (E.I. duPont de Nemours and Co., Inc., Wilmington, Del.) was found to be the most sensitive and rapid method for isolating yeasts and filamentous fungi from blood by Jacques Bille et al. (1983). The detection of fungemia was increased 36.6% by the Isolator, which also was found superior to the Bactec radiometric system (Johnston Laboratories, Cockeysville, MD) by Roy Hopfer, Karen Mills, and Dieter Gröschel (1979).

In addition to improved techniques for the direct demonstration of *Candida* spp. cells in blood cultures, indirect methods were developed for the diagnosis of candidiasis. Immunologically based procedures, such as skin tests and the measurement of serum antibodies, had provided unsatisfactory results as diagnostic tools for *Candida* spp. infections. Investigators began to develop assays to detect candidal antigens or metabolites (Miller, Witwer, Braude, and Davis, 1974; Weiner and Yount, 1976; Kiehn, Bernard, Gold, and Armstrong, 1979; Segal, Berg, Pizzo, and Bennett, 1979). Although these alternatives were promising, they lacked specificity and sensitivity. The efforts in the ensuing years were directed toward the detection in serum of *Candida* D-arabinitol, enolase, and cell wall mannoproteins by more advanced technologic approaches, such as DNA-based methods.

Two approaches were investigated for the immunodiagnosis of aspergillosis. John Schaefer, Bessie Yu, and Donald Armstrong (1976), at the Memorial Sloan-Kettering Cancer Center and Cornell University, found that "biweekly immunodiffusion tests (modified Ouchterlony's test) for *Aspergillus fumiga-*

tus antibodies were helpful for the clinical diagnosis of invasive aspergillosis in a high-risk population" (p. 328). The second approach was the development of tests for the detection of antigens of *A. fumigatus* in serum (Lehmann and Reiss, 1978; Weiner and Coats-Stephen, 1979; Weiner, 1980). Although the first approach is used frequently today for the diagnosis of non-invasive aspergillosis, the second approach seems to have a greater potential for the immunodiagnosis of invasive disease, especially in the immunocompromised host. Negative precipitin tests do not rule out invasive disease. The nature of the antigen detected, a galactomannan, was reported by Errol Reiss and Paul Lehmann (1979), while Lehmann was a postdoctoral fellow with Reiss at the CDC ("Training Trees", Appendix B).

Several commercial kits for the identification of yeasts in the clinical laboratory were developed in the 1970s. At the time conventional methods were tedious and time consuming despite the modifications of the methods of Lynferd Wickerham and Kermit Burton (1948) by Geoffrey Land, E.C. Vinton, G.B. Adcock, and J.M. Hopkins (1975) from the Wadley Institutes of Molecular Medicine in Texas. The first evaluation of the API 20C Yeast Identification System (API Laboratory Products, Ltd., Plainview, NY) was performed in the laboratory of Glenn Roberts at the Mayo Clinic. Roberts, Howard Wang, and Gary Hollick (1976) compared 300 isolates of clinically important yeasts using the API 20C system and conventional methods, and found that the API 20C system was a useful method for yeast identification. A modified version of this system is the test used today in most clinical laboratories; the API computerized database was developed by David Pincus (Questionnaire). This new version was evaluated by Land, B.A. Harrison, K.L. Hulme, Billy Cooper, and J.C. Byrd (1979). Based on Fritz Staib's discovery in 1962 that *C. neoformans* produced melanin when grown on agar containing an extract of *Guizotia abyssinica* seeds, Roy Hopfer and Dieter Gröschel (1975) at the University of Texas M.D. Anderson Cancer Hospital in Houston, and Ira Salkin (1979) at the New York State Laboratory developed a 6-h test that screened for *C. neoformans* colonies.

Paul Standard and Leo Kaufman developed the mycologic exoantigen test between 1976 and 1977. This important advance was accomplished while Standard was conducting research for his public health doctoral dissertation at the University of North Carolina at Chapel Hill under the guidance of Kaufman from the CDC ("Training Trees", Appendix B). By 1975, the identification of the fungal pathogens *B. dermatitidis*, *C. immitis*, and *H. capsulatum* var. *capsulatum* involved the demonstration of their diagnostic morphology in the mycelial form, and the subsequent conversion from the mycelial to the tissue form. Using the exoantigen test, these transformation steps were eliminated because this approach was based on the direct three-day detection of specific soluble antigens from mature cultures of each fungus. The exoantigen method was adapted later for many other fungal pathogens. Commercial exoantigen kits have been available since the 1980s; however, a negative exoantigen test is not always absolute proof that the culture is not *H. capsulatum*. During the 1980s, DNA based technologies (DNA probes) were developed for the rapid identification of these same pathogens.

In the 1970s, the frequent use of coccidioidin (derived from mycelium) for epidemiologic purposes when defining exposed populations, or for the diagnosis of coccidioidomycosis when previous skin tests had been negative, or for the prognosis of this infection in case of anergy to *C. immitis* indicated that improvement in this skin test reagent was needed in certain clinical situations. Hillel Levine, Antonio Gonzalez-Ochoa, and David Ten Eyck (1973), at the Naval Biochemical Research Laboratory, compared the reagent prepared by them in 1969 from cultured spherules with the classic coccidioidin of Charles E. Smith and his co-workers (1940s). Levine et al. (1973) found that spherulin was more sensitive than coccidioidin. Both reagents either together or alone are used today; many approaches to purifying these heterogeneous extracts and other immunoreactive antigens have been attempted.

Host defenses against yeasts

Cryptococcus neoformans

By using both animal models and in vitro assays, preliminary basic research advances in understanding fungal pathogenesis and host-fungus interactions, such as phagocytosis and cell-mediated immunity, were accomplished during the 1970s. By 1974, it had been established that the host's major defense mechanism against *C. neoformans* was cell-mediated and could be measured by detecting delayed hypersensitivity. Juneann Murphy, Jay Gregory, and Howard Larsh (1974), at the University of Oklahoma, Norman, developed their first murine model to study this mechanism in cryptococcosis. During the late 1970s, T-lymphocyte-mediated mechanisms appeared to be the primary means of host defense in this disease in the murine model.

Since the 1970s, Richard Diamond has focused his studies on the host immune mechanisms in opportunistic fungal infections such as aspergillosis, candidiasis, cryptococcosis, and zygomycosis. Diamond and Anthony Allison (1976), at the Michael Reese Hospital in Chicago, established that human peripheral blood leukocytes, excluding T cells, could kill *C. neoformans* in the presence of anticryptococcal antibodies in vitro. "The killing of *C. neoformans* cells appeared to occur by a nonphagocytic mechanism" (Diamond and Allison, 1976, p. 718). Peripheral blood neutrophils, monocytes, and "activated" macrophages had a limited capacity to "ingest" and kill *C. neoformans*. However, ingestion was not essential for the anticryptococcal action of phagocytes. In 1977, Diamond developed a guinea pig model of intraperitoneal cryptococcosis, using intravenous infection to define the relative importance of these killing mechanisms in vivo and the effects of immunosuppression and immunostimulation.

As a result of Glenn Bulmer and M.D. Sans' (1967) finding that cryptococcal polysaccharide inhibits the attachment, but not the ingestion stage of phagocytosis, Thomas Kozel and Terrence McGaw (1979) at the University of Nevada discovered that immunoglobulin G (IgG) is "the principal, if not the sole, opsonin in human serum for phagocytosis of *C. neoformans* by normal mouse macrophages" (p. 260). Opsonization, which facilitates phagocytosis, involved a direct interaction between IgG and the macrophage via an Fc-mediated process.

Candida albicans

For phagocytosis of fungi to occur, phagocytes must come in contact with fungal cells. Richard Diamond and Raymond Krzesicki (1978) showed that human peripheral blood neutrophils could ingest and probably kill *C. albicans* in vitro. They also showed that attachment to neutrophils was inhibited by *Candida* mannans (1 to 10 mg/ml). The neutrophils recog-

nized a protein component on the *Candida* surface which was sensitive to chymotrypsin. On the other hand, the neutrophil receptors for *Candida* cells were susceptible to chymotrypsin as well as trypsin. The damage to *Candida* cells occurred by oxidative mechanisms, which confirmed Robert Lehrer's and Martin Cline's findings in 1969. It is now known that both oxidative and nonoxidative mechanisms are involved in the killing of *C. albicans* by neutrophils and macrophages.

Host defenses against dimorphic fungi

Coccidioides immitis

During the 1970s, Demosthenes Pappagianis (University of California, Davis), who had trained in medical mycology under the guidance of Charles E. Smith, and David Stevens (Santa Clara Valley Medical Center) initiated research on the host immune response in resistance to *C. immitis*. Lovelle Beaman, Pappagianis, and E. Benjamini (1977) reviewed the literature and concluded that although T cells were required for the cell-mediated immune response of bacterial infections, their role in *C. immitis* infections was not well defined. Cell-mediated immunity appeared to be associated with recovery from coccidioidomycosis (skin test positive), and severe infections were regulated by humoral response (high CF titer and skin test negative). Beaman et al. (1977) concluded that "the effector cells that inactivate the spherules and endospores during infection are presumably macrophages and polymorphonuclear phagocytes and that the T-cell population was essential for effective immunity in mice" (p. 585).

John Galgiani, Richard Isenberg, and Stevens (1978) studied, for the first time in fungi, the substances derived from *C. immitis* that cause chemotactic migration of PMNs to the site of infection in vitro. This cellular defense mechanism was mediated by the generation of heat-labile complement-split fragments, but direct leukotactic properties in the absence of serum could be displayed. These results suggested to Galgiani et al. (1978) that alternative and classical complement pathways were activated by mycelial and spherulin filtrates without antibodies being required. Therefore, the inhibition of migration of PMNs by high concentrations of "spherule-derived substances might obstruct efforts to limit *C. immitis* infection" (Galgiani, et al., 1978, p. 864). Galgiani carried out his postdoctoral medical training under the supervision of Stevens ("Training Trees", Appendix B).

Histoplasma capsulatum var. *capsulatum*

By 1977, Dexter Howard at the University of California, Los Angeles had made the following important discoveries on cellular immunity in murine histoplasmosis: macrophages from immunized animals restricted intracellular growth of *H. capsulatum* var. *capsulatum*; the lymphocyte acted as the mediator cell and the macrophage acted as the effector cell; and partially purified lymphocytes "armed" macrophages to suppress the intracellular growth of this fungus. However, supernatants from cultures of immune lymphocytes did not activate macrophages (Howard and Otto, 1977). On the other hand, Ronald Artz and Ward Bullock (1979) at the University of Kentucky studied the suppression activity induced by disseminated histoplasmosis capsulati in mice in order to understand the immune function in humans. They noted a shift of host cell activity or immunoregulatory disturbances from a suppressor (1 to 3 weeks) to a helper mode by week 8, when mice recovered from the infection. These parameters were similar to those observed in humans. Both Howard and Bullock continued their studies on the cellular immunity against *H. capsulatum* through the training and development of graduate and postdoctoral students; Howard retired in the 1990s.

Host defenses against aspergillosis and zygomycosis

Richard Diamond, Raymond Krzesicki, Brad Epstein, and Wellington Jao (1978) demonstrated that neutrophils were also important as host defense mechanisms in invasive aspergillosis and zygomycosis (as mucormycosis), even though hyphae in the lesions were too large to be phagocytized. Light electron microscopic observations showed that neutrophils "attached to and spread over the surfaces of hyphae even in the absence of serum" (Diamond et al., 1978, p. 313). This nonphagocytic mechanism was followed by severe damage and probably death of the hyphae. Damage to the hyphae was inhibited by compounds that affect neutrophil surface functions, motility, and metabolism. The attachment appeared to activate potential fungicidal mechanisms in neutrophils.

Physiology and nutrition

Geoffrey Land, W.C. McDonald, R.L. Stjernholm, and Lorraine Friedman (1975) at Tulane University associated morphogenesis of *C. albicans* with changes in respiration, while Land was conducting research for his doctoral dissertation under the guidance of Friedman. Hyphae "produced more ethanol, evolved less CO_2, and consumed less oxygen than yeasts" (p. 126). This suggested to them that an abrupt change from an aerobic to a fermentative metabolism of glucose occurred. Nickerson's (1954) attempts to describe the morphogenesis of *C. albicans* at the cellular level had prompted the studies by Land and co-workers (1975). Because differences between the predominant cytoplasmic proteins of the two forms were not detected, Marcia Manning and Thomas Mitchell (1980) at Duke University suggested that "proteins unique to each phase may serve a regulatory function" (p. 270).

Bruno Maresca, Gerald Medoff, David Schlessinger, and George Kobayashi (1977) examined further the role of sulphydryl groups in determining the forms of *H. capsulatum* var. *capsulatum* growth (Salvin, 1949; Pine and Peacock, 1958). Maresca et al. (1977) found that 37°C initiated a series of reactions leading to changes in the intracellular level of cyclic AMP. The level of cyclic AMP was five-fold higher in the mycelium form than in the yeast form, "which was an important determinant of the morphological form and its disease-producing potential" (Maresca et al., 1977, p. 447). These investigators also showed that cysteine stimulated oxygen consumption in the yeast but not in the mycelial forms.

Etiology and morphology

Paul Szaniszlo and his students' main research interest at the University of Texas at Austin has been the morphologic, ultrastructural, and more recently the molecular changes associated with the growth of both yeast and hyphal forms of one of the phaeoid (as dematiaceous) fungi that causes phaeohyphomycosis, *Wangiella dermatitidis*. They use this species as a model for most phaeoid pathogens. Szaniszlo, P.H. Hsieh, and J.D. Marlowe (1976) described for the first time the induction of *W. dermatitidis* multi-cellular muriform (as sclerotic bodies) cells in vitro. These bodies appeared to be an intermediate form between yeast and hyphal growth.

Training and education contributions: 1970 to 1979

Brown–Hazen training funds

Particularly acute in the 1970s was the demand for trained clinical laboratory personnel and the need to provide physicians with concise, up-to-date information on fungal pathogens. Several postdoctoral and graduate programs were developed and supported with NIH and Brown–Hazen Program funds. The Brown–Hazen funds came from the royalties set aside from the invention and sale of the antifungal drug nystatin in the 1950s. Initially in 1957, the funds were earmarked for basic research in biochemistry, microbiology, and immunology. By the end of the 1960s, Elizabeth Hazen was disturbed by the lack of medical mycology grant applications. In looking ahead to the expiration of the royalties from nystatin in the late 1970s, the Brown–Hazen Awards Committee decided to focus their efforts on medical mycology (Baldwin, 1981). Training grants were determined to be the best way to further the discipline, and in 1970 the Committee sought out medical mycology training centers and awarded grants to most of them (Research Corporation, 1971). The multiplier effect of these grants to support and develop training centers was a major factor in the development of the discipline during "the years of expansion" in the 1970s.

In 1977, the Dalldorf Postdoctoral Fellowship Award in medical mycology was established to honor Gilbert Dalldorf, the first Committee Chairman and a major force in guiding the work of the Brown-Hazen fund. Dalldorf turned to the Infectious Diseases Society of America (IDSA) to help design a plan for this fellowship program. The Dalldorf Fellowship targeted promising young postdoctoral students, preferably physicians, and provided two years of funding for research in medical mycology. Accepted fellows were chosen by the IDSA Executive Committee and Brown–Hazen Program Committee, based upon their original research ideas, sponsorship, and association with a "strong" medical mycology center (Research Corporation, 1990). Ten fellows, nine physicians and one Ph.D. scientist, received awards before the funds

were depleted in 1991, and many well-known contributors to the discipline have emerged as a result of this unique training grant ("Training Trees", Appendix B). Research contributions from some of these individuals are highlighted under the scientific contributions sections.

National Institute of Health (NIH) training grants

From information gathered through interviews and the questionnaire, initially only two medical centers were awarded NIH training grants that included stipends for students, and six centers received major NIH research funding. The first training award was given to Tulane University in 1958 (to Lorraine Friedman and later to Judith Domer), and it was followed by training and research grants awarded to Washington University from 1976 to 1992 (George Kobayashi, P.I. for training and Gerald Medoff, P.I. for research), the University of California, Los Angeles (Dexter Howard), and Temple University (Helen Buckley). Major NIH funding support as Program Project Grants for basic research studies was granted to the University of Oklahoma, Norman (Howard Larsh) and the University of Cincinnati (Ward Bullock). H. Jean Shadomy and John P. Utz were awarded a training grant (MCV, VCU) between 1969 and 1970. Judith Domer stated in 1994 that these extramural NIH grants, as well as the intramural awards for medical mycology studies at the NIH and CDC provided important support for training and education programs until the late 1980s and the early 1990s. Figure 95 locates the medical mycology training and research programs established up to 1979 in the United States.

University of Kentucky

Kentucky is in the center of an endemic area for blastomycosis and histoplasmosis capsulati. Based on a feasibility study, a medical mycology program was established at the University of Kentucky and a mycology research laboratory at the VA Hospital in Lexington in August 1969. Michael Furcolow and Ernest Chick were the guiding forces and codirectors behind this effort and their research contributions are noted under the scientific contributions sections. Norman Goodman, a student of Howard Larsh ("Training Trees", Appendix B), was recruited in 1971 to direct the training program which continued until 1984 when it was closed (Goodman,

1985). During the first year, this dual program received eight grants from various agencies totaling $230 000.

More than 1000 individuals were trained at the University of Kentucky between 1971 and 1984 with continuing education short-term and long-term courses ranging from one month to one week each year. Initially two courses were offered and in 1973 the program was expanded to four courses and a two-day clinical conference. The July course included the study of all pathogenic fungi; the one-week September course was dedicated to the systemic pathogens; the one-week December course covered the dermatophytes; and the three-day course was dedicated to the serology of the systemic fungi. Howard Larsh and Norman Conant came as regular faculty members for periods of up to two weeks (Chick and Furcolow, 1971; N. Goodman, personal communication, May 24, 1994).

From 1972 to 1977, Brown–Hazen Foundation grants were awarded to Michael Furcolow and Ernest Chick, and later to Furcolow and Norman Goodman to support this comprehensive training program in medical mycology (Research Corporation Quarterly Bulletins, 1971 and 1975). The training grants helped to defray the costs to medical technologists who had limited budgets and would not have been able to attend without financial assistance (N. Goodman, personal communication, March 14, 1994). The primary target groups for training at the University of Kentucky were practicing physicians and laboratory personnel doing day-to-day diagnostic bench work. Since 1975, Goodman directed the research of 20 microbiology students, who carried out medical mycologic studies for their dissertations. Nine postdoctoral students, including Glenn Roberts, who is currently directing the mycology diagnostic laboratory at the Mayo Clinic, also had received training at Goodman's laboratory ("Training Trees", Appendix B). Roberts came to Kentucky as Goodman's postdoctoral fellow in 1972 after receiving his Ph.D. degree with Larsh at the University of Oklahoma.

Temple University School of Medicine

Fritz Blank applied again in 1970 and was awarded a Brown–Hazen grant of $240 000 to develop a broad training and research program over a period of six years at the Department of Dermatology at Temple University (Baldwin, 1981). The goal of this pro-gram was to offer medical mycologic training to qualified predoctoral students obtaining M.S. and Ph.D. degrees, as well as postdoctoral individuals who had a background in microbiology, biochemistry, or organic chemistry. Several well-known investigators, including Carl Abramson (Ph.D.), Richard Calderone, and Roy Hopfer (postdoctoral), received medical mycology training under Blank's guidance ("Training trees", Appendix B). Between 1972 and 1976, Blank, Sarah Grappel, and R.J. Yu (in 1976) offered three, five-day courses (every two years), primarily for physicians and microbiologists interested in the science of medical mycology and its recent advances. This course emphasized the clinical aspects of various mycoses and included an introduction to general mycology, as well as a comprehensive review of the fungi that were known at the time to cause human and animal disease. Up to 18 contemporary medical mycology leaders participated as faculty members or guest speakers for these three courses. The medical mycology program initiated by Blank at Temple University was terminated after his death on April 22, 1977. Grappel moved the same year to the Smith, Kline, and French Laboratories in Philadelphia and Helen Buckley moved to Temple as an Associate Professor of Microbiology and Director of the Mycology Laboratory at the University Hospital. Buckley continued to teach ("Training trees", Appendix B) and conduct research in medical mycology at Temple until her death. Her interests were the pathogenesis of *C. albicans*.

University of California at San Francisco

In 1971, Carlyn Halde organized and began offering workshops in medical mycology at UCSF. These workshops brought in well-known medical mycologists as faculty and were established to support continuing education for laboratory personnel and interested physicians, especially pathologists. Along with Halde, important contributors were Michael McGinnis, Leanor Haley, Michael Rinaldi, and Miriam Valesco, who taught basic medical mycology and kept participants up to date on developments in the field (C. Halde, personal communication, December 9, 1993; Halde, Course Manuals, 1971 and 1987). For 27 years, Halde also contributed the chapter entitled, Infectious Diseases: Mycotic, of the book, *Current Medical Diagnosis and Treatment*, that has been published annually since 1962.

Washington University

At Washington University, St. Louis, Missouri, the medical mycology training and research program started in 1972, when the basic scientist George Kobayashi obtained a one-year Brown–Hazen training grant. He was trained by Lorraine Friedman ("Training Trees", Appendix B). The Brown–Hazen grant was subsequently renewed for several years (1973 to 1976) and these funds enabled this program to secure three five-year training grants from the NIH that began in 1976 and ended in 1992. The program funded four students, including stipends and was renewed twice. The principal investigators of the program were Kobayashi for training and the physician Gerald Medoff for research. The research component of the program was divided into four major sections that encompassed four aspects of the discipline of medical mycology: the molecular biology and dimorphic nature of *H. capsulatum* var. *capsulatum*, the development of new and rapid laboratory diagnostic methods, the study of immunological responses and host resistance to fungal infection, and the investigation of new antifungal therapies (see scientific contributions sections).

The Washington University program also provided intensive multi-disciplinary training for a small group of postdoctoral Ph.D. or M.D. students. The goals of the program were to stimulate and support sophisticated research in medical mycology with the objective of equipping these students for faculty positions at other universities, where they would conduct medical mycologic activities. These postdoctoral students were exposed to medical mycology, molecular biology, immunology, and biochemistry in addition to clinical and diagnostic training opportunities (Kobayashi, 1982; personal communication, August 26, 1994). An important strength of this program was that it enabled students to work with the faculty members in the Divisions of Dermatology and Infectious Diseases, the Department of Microbiology and four hospitals in St. Louis, including Barnes Hospital. The program emphasized that academic course work for students would be closely associated with patient care. Each postdoctoral student ("Training Trees", Appendix B) benefited from at least two years of individual attention and work assignments that included time in the clinical laboratories. A number of important contributors to basic research in the discipline emerged from this program; their studies are summarized under the scientific contributions sections. In addition to important research contributions, most of the individuals who completed their training in this program were able to obtain independent funding to continue their careers at other centers. (G. Kobayashi, personal communication, September 1, 1994). In 1979, the NIH designated this institution as one of the medical mycology centers in the United States.

Medical College of Virginia/VCU (MCV, VCU)

The physician Eric Jacobson, who trained in medical mycology under Gerald Medoff and George Kobayashi at Washington University, became a joint faculty of the Veterans Hospital and the Division of Infectious Diseases at MCV, VCU in 1974 to conduct basic research on *C. neoformans*. Since the 1970s, Jacobson has provided guidance to several graduate and postdoctoral students ("Training Trees", Appendix B). In the 1970s, the training and research climate at MCV, VCU was sufficiently strong for the award of the first Dalldorf Fellowship in Medical Mycology to the physician, Thomas Kerkering, in 1978. Kerkering studied with Jean and Smith Shadomy and Ana Espinel-Ingroff, who had been hired to conduct laboratory research in 1971. The latter is nationally and internationally recognized for her work on the standardization of antifungal susceptibility tests for yeasts and moulds. Kerkering, a fellow in the Division of Infectious Diseases, concentrated his research on the development of a simple diagnostic test for the rapid and accurate diagnosis of *C. albicans* infections. Among the formal students and mentees of the Shadomys ("Training Trees", Appendix B), several well-known contributors to the field of medical mycology are Robert Fromtling at Merck Laboratories and Dennis Dixon, the Chief of the Bacteriology and Mycology Branch at the NIH in 1996.

The Centers for Disease Control (CDC)

Errol Reiss came to the CDC in 1974. Reiss received his Ph.D. degree at Rutgers University under the direction of Walter J. Nickerson, a Professor of Microbial Biochemistry, and his medical mycology postdoctoral training at the NIH at the laboratory of Herbert F. Hasenclever, Chief of the Mycology Section at that time (E. Reiss, personal communication, January 25, 1994). Since the 1970s, several individuals have received medical mycology post-doctoral training at Reiss' laboratory ("Training Trees", Appendix B). In addition to his scientific contributions, Reiss published a book in the field of immunochemistry of mycoses caused by *Aspergillus, Candida, Cryptococcus,* and *Histoplasma.*

During "the years of expansion", Leanor Haley modified her training curriculum to respond to the increased prevalence of opportunistic fungal diseases and advances in medical technology. In her first years at the CDC (late 1960s), transplantation became important and few people knew how to interpret the data coming from the clinical laboratory. The CDC immediately focused on training physicians and laboratory personnel to identify and understand the new and emerging opportunistic fungi. When the AIDS epidemic began in the 1980s, Haley had to adapt her training again. Safety became an important issue and all clinical laboratories personnel had to learn new procedures to protect the patients and themselves (L. Haley, personal communication, December 31, 1993).

Libero Ajello co-authored an atlas on the histopathology of fungi with F.W. Chandler and William Kaplan in 1980. It became a prototype for many subsequent training efforts. The *Color Atlas and Text of the Histopathology of Mycotic Diseases* (Appendix C) included a vast selection of photomicrographs that represented fungal lesions with corresponding captions to facilitate the rapid review or differential diagnosis of fungi in tissue. Its audience was pathologists, clinicians, and laboratory personnel in need of an accurate, yet fast diagnostic textbook for fungal infections. Thousands of people, mostly physicians and clinical laboratory personnel, were trained via this reference book.

The National Institutes of Health (NIH)

In 1971, John Bennett became the Head of the Clinical Mycology section. In 1974, the medical mycology section was merged into the Laboratory of Clinical Investigation with Kyung Joo Kwon-Chung as a leading researcher. In 1977, Kwon-Chung began to make her molecular-based contributions to the third edition of the textbook, *Medical Mycology,* that Chester Emmons had published for the first time in 1963 with John Utz and Chapman Binford. Bennett and Kwon-Chung continue the tradition of Emmons, Utz, and Herbert Hasenclever and others as leaders in clinical and basic research

and provide important informal training opportunities in medical mycology at the NIH. Their contributions are noted under the scientific contributions sections. Important investigators, including two Dalldorf Fellows, who received doctoral or postdoctoral training under Bennett and Kwon-Chung are the physician John Rex (under Bennett) and the Ph.D. scientists, Paul T. Magee, Julie Rhodes, William Whelan, and Brian Wickes ("Training Trees", Appendix B).

Duke University

When Conant retired in 1974, Thomas Mitchell, a graduate of the Tulane University program under Lorraine Friedman ("Training Trees", Appendix B), was hired to continue the mycology program. As the new Director of the Medical Mycology Laboratory and Assistant Professor of Mycology, Mitchell assumed the responsibility for teaching medical mycology in the Medical and Microbiology Graduate Schools. He established a basic research program that has been supported by NIH extramural funds. In 1975, he also reorganized and continued Conant's renowned Duke Mycology Summer Course and has offered this continuing education program since that time. In Mitchell's laboratory several undergraduate, graduate, medical, and postdoctoral students have received medical mycology training and conducted related research projects under his direction ("Training Trees", Appendix B). In 1977, D. Durack became the Chief of the Infectious Diseases Division and recruited the physician, John Perfect, who became involved in the training component of this program. John Perfect and Harry Gallis, another infectious diseases specialist, assisted as clinical faculty members in teaching the Duke Summer Course.

Georgetown University

In 1974, Richard Calderone began his academic career at Georgetown University as Assistant Professor in the Department of Microbiology. He completed his medical mycology postdoctoral studies at Temple University under the direction of Fritz Blank and Sarah Grappel. Since 1978, Calderone has been awarded NIH grants for his studies on the pathogenicity of *C. albicans.* Several individuals have obtained medical mycology training at Calderone's laboratory ("Training Trees", Appendix B).

98

The Johns Hopkins University

In 1974, William Merz came to Johns Hopkins Hospital from Columbia Unversity, where he had trained under Margarita Silva-Hutner ("Training Trees", Appendix B), to establish and direct a medical mycology service laboratory. He had a joint appointment as an Assistant Professor in the Department of Pathology at the University. Merz focused his training efforts on laboratory personnel who were establishing or working in medical mycology service laboratories and for physicians doing postdoctoral training in medical mycology. Since 1975, he has offered a medical mycology course for the same audience (W. Merz, personal communication, January 5, 1994).

Montana State University

Jim Cutler established the first medical mycology research laboratory at this university in 1974. He obtained his M.S. degree under the direction of F. Swatek, a student of Orda Plunkett at the California State University, Long Beach, and his Ph.D. at Tulane University under the guidance of Lorraine Friedman ("Training Trees", Appendix B). Currently, Cutler is directing perhaps the only NIH-funded medical mycology graduate program at Montana State University (See Chapter VI)

University of North Carolina (UNC) at Chapel Hill

Michael McGinnis trained under the classical mycologist Lois Tiffany in the Department of Botany at Iowa State University, and went to CDC for his postdoctoral training under the guidance of Libero Ajello and Leanor Haley ("Training Trees", Appendix B). During his years at UNC (1975-1988), he taught a four-credit medical mycology course each year in the Department of Parasitology and Laboratory Practice (McGinnis, 1994). The course usually had 12 to 15 students and consisted of laboratory sessions and lectures. He also taught medical technology students, began the API workshop series with Richard D'Amato, and expanded the ASM Annual Meeting and Travel Workshops in the fungal identification of moulds and yeasts (McGinnis, personal communication September 12, 1994). He is well-known for his *Laboratory Handbook of Medical Mycology* (Appendix C) published in 1980. This handbook became an excellent

training and diagnostic resource in the laboratory for helping to clarify the nomenclature of the mycoses, especially infections caused by the phaeoid fungi (see scientific contributions); for introducing contemporary terminology to medical mycology (e.g., conidium versus spore); and for clarifying the taxonomy of certain fungi. When McGinnis became editor of the *Clinical Microbiology Journal* (1981–1990), replacing Lorraine Friedman, he expanded the scope of the journal to include the publication of more medical mycology papers, including new discoveries. McGinnis also created the review series called *"Current Topics in Medical Mycology"*. He was the Chief Editor of the *Journal of Medical and Veterinary Mycology* (known now as, *Medical Mycology*).

University of Texas Health Sciences Center (UTHSC) at San Antonio

The history of medical mycology at UTHSC began when the physician David Drutz arrived there as Chief of Infectious Diseases in 1975. He brought with him an interest in the study of fungal infections. He recruited the physicians Richard Graybill and Marc Weiner and the Ph.D. scientist Milton Huppert. These individuals became involved in the development of animal models to evaluate fungal therapy (Graybill) and serologic tests for the diagnosis of aspergillosis (Weiner) and coccidioidomycosis (Weiner and Huppert). Since the 1970s, informal medical mycology training has been provided by these medical mycologists and other investigators that joined the group at UTHSC in the 1980s and 1990s ("Training Trees", Appendix B). Huppert had been Chief of the Mycology Research Laboratory at the VA Hospital in Long Beach, California and was offered the same position at the Audie Murphy Memorial VA Medical Center in San Antonio. Huppert received his Ph.D. degree from Columbia University under the tutelage of Rhoda Benham ("Training Trees", Appendix B). After his years at Columbia, he organized a medical mycology course for graduate students at the University of North Carolina. Huppert died of lung cancer on January 31, 1984.

Baylor University

An important advocate and promoter of medical mycology during the 1970s and 1980s was Bill

Cooper, who also trained under Lorraine Friedman ("Training Trees", Appendix B). Cooper was baptized "Billy" and "he was best remembered and loved by the 'rank and file' for his ability to teach diagnostic medical mycology painlessly, lovingly and always with a gleam in his eye" (Domer and Friedman, 1988). Cooper was an outstanding educator, who took his ASM Traveling Workshops across the country at least 25 times from 1974 to 1986 to teach practical approaches in the laboratory for the proper identification of fungal pathogens. In addition, Cooper was well known for his participation in numerous Continuing Education Workshops and seminars during his career. He died on August 6, 1987 at the age of 51, prematurely ending his important career and contributions to the development of the discipline of medical mycology (Domer and Friedman, 1988). He was not replaced.

Methodist Medical Center, Dallas

Geoffrey Land obtained his Ph.D. at Tulane University under the guidance of Lorraine Friedman and was a postdoctoral fellow at Duke University. Land has served as a faculty member for more than 200 workshops and has formally taught medical mycology to graduate and medical students. He also directed seven research dissertations at the Methodist Medical Center in Dallas ("Training Trees", Appendix B). Land collaborated with Billy Cooper, his mentor, in developing the concept of the traveling workshop during the 1970s.

The Mayo Clinic

Glenn Roberts, who trained under Howard Larsh, Norman Goodman ("Training Trees", Appendix B), and the physician Elmer Koneman, Head of the Microbiology Laboratories at the VA Hospital, Denver, Colorado, gave two one-week medical mycology workshops for approximately 100 participants each year. The laboratory diagnosis of fungal infections was the primary emphasis of these workshops that have been directed towards clinical laboratory personnel involved in the recovery and identification of fungi from clinical specimens. Workshops currently are offered in response to demand and have been structured accordingly (L. Stockman, personal communication, June 6, 1994). Another important feature of medical mycology training at the Mayo Clinic are the residency and fellowship programs for postdoctoral students from the medical and dental schools (Mayo Clinic Brochure, undated).

Harvard School of Public Health

Charlotte Campbell and Martha Berliner were awarded both NIH research grants (G. Kobayashi, personal communication, September 1, 1994) and Brown–Hazen training grants (Baldwin, 1981). As a team, they taught medical mycologic courses and seminars for public health students and short clinical lectures around the Boston area from the mid-1960s to the late 1970s. Berliner obtained her doctoral degree under the guidance of the general mycologist, Lindsay Olive, while he was a professor of botany at Columbia University. Olive had trained under W.C. Coker at the University of North Carolina in Chapel Hill.

Boston University

In the late 1970s, Richard Diamond established a medical mycologic research laboratory at Boston University Medical Center. Between 1983 and 1989, the physicians, Edwin Smail, Stuart Levitz, and David Stein were awarded Dalldorf Fellowships to pursue their postdoctoral training with Diamond (Research Corporation, 1990; "Training Trees", Appendix B).

University of California, Davis

Since the 1970s, Demosthenes Pappagianis has maintained an active medical mycology graduate and postdoctoral (M.D. and Ph.D. graduates) training program at this institution. Pappagianis trained under Charles E. Smith at the University of California at Berkeley. Another eminent medical mycologist at the University of California, Davis, was the physician Paul Hoeprich, who trained Michael Rinaldi ("Training Trees", Appendix B). Rinaldi went to the University of Texas in San Antonio, after a short stay at Montana State University, to establish a reference medical mycology laboratory in the 1980s.

Santa Clara Valley Medical Center

Medical mycology postdoctoral fellowships have been offered since the early 1970s at Santa Clara

Valley Medical Center in the laboratory of the physician David Stevens. The principal interest at Stevens' laboratory has been the immunology and chemotherapy of infectious diseases. Stevens finished his postdoctoral training at Stanford University, Palo Alto, California (D. Stevens, personal communication, June 14, 1996). The physicians John Galgiani, Alan Sugar, and David Denning and the Ph.D. scientists Karl Clemons and Christine Morrison are among the individuals who received medical mycology training under the guidance of Stevens ("Training Trees", Appendix B). After training with Stevens, Galgiani established a medical mycology laboratory at the University of Arizona, Sugar at Boston University and Denning in England. Stevens, Galgiani, and Sugar are active members of the MSG group and have participated in

several clinical trials conducted by this group (see scientific contributions sections); Stevens is the Chair of the clinical trials for aspergillosis and Galgiani for coccidioidomycosis. The Ph.D. scientist, Elmer Brummer, joined Stevens in 1978 as a Research Associate. Brummer has focused his research mostly on host-fungal pathogen interactions, macrophage and PMN antifungal mechanisms and the effects of the activation of these phagocytic cells by IFN-γ for enhancement of the killing of several fungal pathogens. Brummer obtained his medical mycologic postdoctoral training with Rebecca A. Cox at the San Antonio Chest Hospital, San Antonio, TX. The individuals who have trained under Brummer's direction are listed in Appendix B ("Training Trees").

VI. The era of transition: 1980 to 1996

Figure 96. John Perfect (p. 113)

Figure 97. Juneann Murphy and her student Paul Fidel (pp. 114-115) (Courtesy of Dr. Fidel)

Figure 98. Stuart M. Levitz and his laboratory team (left to right): Lauren Yauch, Charlie Specht, Jen Lam, Levitz, Salamatu Mambula, Michael Mansour, Keya Sau, Jianmin Chen, and Shu-hua Nong (p. 118) (Courtesy of Dr. Levitz)

Figure 99. Paul Szaniszlo (2nd left) (pp. 93-94, 122), Gary Cole (1st left) (pp. 116-117, 119), and graduate students Zheng Wang, Wen Chen and his wife, Susan, former post-doc Ming Peng and grad student Xiangcang Ye (Courtesy of Dr. Szaniszlo)

Figure 100. John Nathaniel Couch (1896–1986) (p. 122)
(Courtesy of Dr. Szaniszlo)

Figure 101. Dennis Dixon (p. 122)

Figure 102. NIH-funded Predoctoral Medical Mycology Program at Montana University, under Jim Cutler (pp. 122-123) (Courtesy of Dr. Cutler)

Figure 102a. Donald Greer, Morris F. Shaffer, Judith Domer (p. 124), George Kobayashi (first row), Thomas Mitchell (second row, 1st), and Angela Restrepo (second row, 3rd) (Courtesy D. Restrepo)

Figure 103. Judith Rhodes while she was finishing her postdoctoral training at NIH (p. 125)

Figure 104. Michael Rinaldi (p. 125) (Courtesy of Imedex)

Figure 105. Elias Anaissie (p. 125) (Courtesy of Imedex)

Overview

In the early 1980s, the first cases of acquired immunodeficiency syndrome (AIDS) were diagnosed. It became evident that these were not isolated cases but that the world was dealing with a pandemic. Unfortunately, as expressed by 15.1% of the questionnaire respondents (Table 3, Appendix A), AIDS and the induction of immunosuppression during cancer and transplantation chemotherapy became important events for the development of medical mycology. Opportunistic and classic fungal pathogens cause severe and fatal disease in increasing numbers in AIDS, cancer, and transplantation patients and the field of medical mycology can no longer be neglected. Difficulty in the management of opportunistic infections in these patients spurred the development of new therapies, laboratory diagnostic tests, and initiated studies of pathogenesis and immunology.

Although basic research studies had begun earlier, researchers became more specialized during the 1980s and focused their studies on specific aspects of a fungal disease or its etiologic agent. A great deal of research on the molecular biology, cellular immunity, host-parasite interactions, genetics, and the immunochemistry of fungi was initiated by established and new investigators. Most of these studies were performed by graduate and postdoctoral students under the guidance of their major professors, sponsors, and in collaboration with investigators at other institutions. During this period, specialists in infectious diseases and other areas of microbiology became interested in the basic and applied study of fungi, e.g., basic geneticists, immunologists, immunochemists, molecular biologists, and classic mycologists. The genetics of fungi, especially of *C. albicans, C. neoformans,* and *H. capsulatum* var. *capsulatum* were the subject of considerable study including DNA mapping, transformation/recombination, and gene cloning. The introduction of molecular biology and genetics was considered an extremely important advance for the discipline by 15.7% of the questionnaire respondents (Table 4, Appendix A). These individuals believe that basic research in these areas contributed to a clearer understanding of all aspects of the field, ranging from the properties involved in the virulence of a fungus to its interactions with host defenses.

The role of phagocytes, lymphokines, and defensins in the destruction of the systemic and opportunistic fungal pathogens was studied by several groups using animal models. In addition, it was demonstrated that fungal antigens induced lymphoid cells that could either amplify or suppress the immune response in a regulatory fashion. Because of this, the immunochemistry of several fungal antigens was investigated. Preliminary data demonstrated the potential role of colony stimulating factors as useful adjuncts to antifungal therapy for aspergillosis in immunocompromised patients. The C3d receptors for C3 complement fragments of *C. albicans* were identified and purified. Several fungal enzymes were postulated as important virulence factors. This was followed by their purification, characterization, and gene cloning. The proteinases of *C. albicans* were the enzymes most intensively studied.

The development of more accurate and rapid laboratory diagnostic tests was accomplished by the application of DNA-based methodology and other numerous technological advances. DNA-based procedures were employed as reliable epidemiologic tools and the taxonomic issues of some dermatophytes and phaeoid fungi were clarified.

Two collaborative groups were singled out as important contributors to the development of medical mycology in the United States by the questionnaire respondents, the NIAID-MSG and the NCCLS Subcommittee for Antifungal Susceptibility Tests (Table 3, Appendix A). The formation of these groups was vital to adequately evaluate antifungal therapy and to provide standards for in vitro testing. The collaborative concept was then applied to clinical and applied research investigations. The NIAID-MSG collaborative, randomized, prospective clinical trials established better therapeutic regimens for the management of certain systemic diseases in patients with or without AIDS. The NCCLS Subcommittee, established in 1982, proposed standard testing conditions for yeast and mould testing (NCCLS M27-A and M38-A documents).

Although financial support for medical mycology research was increased for a few years, there was a substantial reduction of federal funds and a depletion of private sources by the end of the 1980s. Basic and applied research has been funded mostly by NIH grants and applied research has pharmaceutical and corporate support programs.

From information gathered through the interviews and the review of scientific papers published since 1980, a major transition in medical mycology leadership was noted. Many leaders of applied research and training programs began to retire, but were not replaced in many institutions because of lack or loss of financial support, especially for training. As a result, established and formal training programs were closed or greatly reduced in scope and medical mycology courses as complete units also nearly ceased to exist, which created a crisis in medical mycology training.

By the end of 1996, the following medical mycologists had either changed roles, retired or died, and their medical mycology activities were discontinued: Everett Beneke and Alvin Rogers at Michigan State University (replaced by Leonel Mendoza); Libero Ajello, Lucille Georg (died), William Kaplan, and Leanor Haley (died) at the CDC; Glenn Bulmer, George Cozad (died), Howard Larsh (died), and H. Muchmore at the University of Oklahoma; Richard Duma, Jean Shadomy, Smith Shadomy (died) and John Utz at the Medical College of Virginia/VCU; Morris Gordon, Dennis Dixon and Ira Salkin at the New York State Department of Health; Carlyn Halde at the University of California, San Francisco; Hillel Levine at the Naval Biological Laboratories; John Rippon at the University of Chicago; Margarita Silva-Hutner and Irene Weitzman at Columbia University; Dexter Howard at the University of California at Los Angeles, and James Sinski at the University of Arizona. Other prominent medical mycology leaders died during the Era of Transition: Charlotte Campbell, Norman Conant, Bill Cooper, Chester Emmons, and Milton Huppert.

Based on reviews of the literature and interviews, the following investigators became active in the field at recognized centers conducting medical mycology research or training: Sandra Bragg, Mary Brandt, Brent Lasker, Timothy Lott, and Christine Morrison at the CDC; Wieland Meyer and Rytas Vilgalys at Duke University; Jeffrey Edman at the University of California, San Francisco; David Schwartz and Charles Cantor at Columbia University; Gilbert Chu, Ronald Davis, and Douglas Vollrath at Stanford University; Mahmoud Ghannoum at Cleveland State University; William Powderly, D. Schlessinger, and W. Goldman at Washington University; Elsa Wadsworth and Lisa Linehan at Georgetown University; Garry Cole at the University of Texas, Austin (currently at the Medical College of Ohio); Patricia Sharkey at the University of Texas Health Sciences Center, San Antonio; P. Johnson and John

Rex at the University of Texas Health Science Center, Houston; Brian Morrow at the University of Iowa; Leslie Stockman at the Mayo Clinic; Nancy Hall at the University of Oklahoma, Oklahoma City; Timothy Buchman, Patricia Charade, Michelle Rossier, John Wingard, and Rein Saral at the Johns Hopkins Medical School.

Among other centers where medical mycology research or training is being conducted are: the University of Minnesota Medical School by Stewart Scherer; San Antonio Chest Hospital by Rebecca Cox; University of South Florida, College of Medicine, by Julie Djeu; Wayne State University School of Medicine by J. Sobel and Jose Vazquez; Indiana University School of Medicine, by Marilyn Bartlett and Lawrence Wheat; Medical College of Ohio by Paul Lehmann; State University of New York by Eric Spitzer and Silvia Spitzer; Albert Einstein College of Medicine by Arturo Casadevall and B. Currie; Medical College of Georgia by John Fisher and C. Newman. In addition, the listing of the NIAID-MSG members in April 1996 included 60 physicians, who were involved in the clinical trials conducted by this group.

1980 to 1996

AIDS and opportunistic candidal infections

In 1981, the CDC's Morbidity and Mortality Weekly Report described the first AIDS cases and by 1984 3000 cases were reported. From 1981 to 1984, the diagnosis of AIDS was clinical, being "based on the finding of a severe opportunistic infection suggestive of a defect in cell-mediated immunity or on the finding of an unusual malignant process in a previously healthy person or both" (Klein et al., 1984, p. 354). By 1983, the incidence of candidiasis, including oral infections, was noted in many AIDS patients. The presence of an unexplained oral *Candida* spp. infection in individuals with a high risk for AIDS became a predictor of the development of serious opportunistic infections and AIDS.

John Wingard et al. (1991) reported that in the spring of 1990 the replacement of intravenous miconazole by oral fluconazole for antifungal prophylaxis in marrow transplant recipients had created an increase in both colonization and disseminated infections caused by *C. krusei* at the Johns Hopkins Oncology Center. During the ensuing years, reports continued to appear in the literature regarding the incidence of non-*C. albicans* as etiologic agents of severe disease among immunocompromised patients in other institutions.

Antifungal therapy

Cryptococcosis

From January 1966 through June 1980 the medical records of 41 patients diagnosed with cryptococcosis at MCV, VCU were reviewed by Thomas Kerkering, Richard Duma, and Smith Shadomy (1981). They observed that pulmonary cryptococcosis was rarely included in the differential diagnosis, which resulted in missed diagnoses and therapeutic errors. Since the evolution of untreated pulmonary disease was dissemination in immunocompromised patients, these patients were treated with antifungal therapy.

In 1987, William Dismukes and the members of the NIAID-MSG tested the hypothesis that the duration of the combined flucytosine-amphotericin B therapy for *C. neoformans* meningitis, established by John Bennett et al. in 1979, "might be shortened further from six to four weeks, thus reducing toxicity without compromising efficacy" (p. 335). This clinical trial was a multicenter, prospective, randomized effort that enrolled 194 patients with cryptococccal meningitis. The study results demonstrated that a four-week regimen should be reserved for patients who have meningitis without neurologic complications, underlying disease, or immunosuppressive therapy. Patients who did not meet these criteria should have six weeks of therapy. Alan Stamm et al. (1987) also reported several side effects of this combined regimen.

It became evident that the established treatment regimen of combined amphotericin B and 5-FC was unsuccessful in AIDS patients with cryptococcal meningitis (75 to 85% versus 40 to 50% success rate); up to 10% of patients were diagnosed with this infection. Further, after treatment was ended most patients relapsed. During the early 1980s, two new oral antifungals were developed, fluconazole by Roerig Pfizer in England and itraconazle by Janssen Pharmaceutica in Belgium. The first reports demonstrated that these two antifungals could be alternatives to lifelong amphotericin B treatment against *C. neoformans* meningitis in AIDS patients (Stern et al.,

1988; Sugar and Saunders, 1988; Denning, Tucker, Hanson, Hamilton, and Stevens, 1989). In 1991, David Stevens, Ira Greene, and Onnie Lang, at the Santa Clara Valley Medical Center, reported that fluconazole therapy could also prevent thrush in AIDS patients. However, it was noted that only controlled trials with large numbers of patients could establish the efficacy of these antifungal agents. In the same year, three collaborative trials were conducted to evaluate fluconazole, one by the California Collaborative Treatment Group (Bozzette et al., 1991) and two by the NIAID-MSG and the AIDS Clinical Trials Group (Saag et al., 1992; Powderly et al., 1992). In 1991, Samuel Bozzette et al. evaluated the efficacy of maintenance therapy with fluconazole in 84 AIDS patients who had completed successful primary therapy for *C. neoformans* meningitis. They compared fluconazole with placebo and found a 37% relapse rate among patients assigned to the placebo versus a 3% relapse rate among the patients receiving oral fluconazole. These investigators concluded that in patients with AIDS, "silent persistent infection is common after clinically successful treatment" (p. 580) for this disease. Oral fluconazole maintenance therapy was highly effective in preventing recurrent infections.

On the other hand, William Powderly et al. (1992) compared fluconazole maintenance therapy versus amphotericin B after primary treatment of 218 AIDS patients with *C. neoformans* meningitis. Oral fluconazole was superior (97% relapse free) to amphotercin B (78% relapse free). In the same year, the MSG Group compared intravenous amphotericin B with oral fluconazole as primary therapy for acute cryptococcal meningitis in AIDS patients (Saag et al., 1992). Treatment was successful in 40% of 63 patients receiving amphotericin B and in 34% of the 131 fluconazole recipients. The difference between the groups in overall mortality (18% on fluconazole versus 14% on amphotericin B) was not significant. Their conclusion was that, "although fluconazole was an effective alternative to amphotericin B as primary treatment for patients who were at low risk for treatment failure, the optimal therapy for patients at high risk (abnormal mental status) remained to be determined" (Saag et al., 1992. p. 83). Other primary therapeutic approaches that have been advocated include the use of higher doses of fluconazole and itraconazole and combined flucytosine and fluconazole therapy. The NIAID-MSG also conducted a trial consisting of a two-week induction phase with either amphotericin B alone or combined with flucytosine, followed by eight-weeks of either fluconazole (400 mg/day) or itraconazole (200 mg twice a day). Preliminary data showed no significant difference between the two regimens (Van der Horst, 1997).

Blastomycosis and histoplasmosis capsulati

Oral ketoconazole (developed in the 1970s by Janssen Pharmaceutica in Belgium) was approved by the Food and Drug Administration (FDA) in 1981. When ketoconazole was evaluated during its phase II clinical trials, preliminary data had shown encouraging results in patients with histoplasmosis capsulati, blastomycosis, non-meningeal cryptococcosis, and coccidioidomycosis (NIAID-MSG, 1985). Because of that, a prospective, multicenter, randomized clinical trial was conducted by these members (1985) to compare the efficacy and toxicity of a low-dose regimen (400 mg/day) of ketoconazole with a high-dose regimen (800 mg/day) in 80 patients with blastomycosis and 54 with histoplasmosis. Ketoconazole was effective in 85% of immunocompetent patients with non-life-threatening, nonmeningeal forms of these infections. But, high dosages were associated with severe side effects (NIAID-MSG, 1985).

By 1992, the success rates for amphotericin B and ketoconazole therapies ranged from 70 to 95% for blastomycosis and histoplasmosis capsulati. In 1992, the efficacy and toxicity of orally administered itraconazole in the treatment of nonmeningeal, non-life-threatening forms of blastomycosis and histoplasmosis was investigated by William Dismukes and the NIAID-MSG members. They conducted a multicenter trial (14 university centers) and evaluated 48 patients with blastomycosis and 37 with histoplasmosis diagnosed by culture or histopathologic evidence. A 90 to 95% success rate was reported among blastomycosis patients and an 81 to 86% success rate among histoplasmosis capsulati patients. Itraconazole became an alternative to either amphotericin B or ketoconazole for the treatment of these infections. However, Peter Pappas et al.'s review of the medical records of 185 patients (between 1956 and 1991) with blastomycosis at the University of Alabama and affiliated hospitals revealed a marked increase (24% since 1981) in the number of immunocompromised patients with blastomycosis. Because the disease appeared to be more aggressive in immunocompromised patients than in normal patients, early and intensive amphotericin B therapy was required instead of the alternative oral agents (Pappas et al., 1993). This mycosis had been the least often associated with immune disorders.

In the United States and Latin America, histoplasmosis capsulati was found among 2 to 5% of patients with AIDS who resided in areas endemic for *H. capsulatum* var. *capsulatum* (Wheat et al., 1991). The prevalence was as high as 25% in the areas of Kansas City and Indianapolis. The relapse rate was 50% in patients treated with ketoconazole and 10 to 20% in those receiving amphotericin B maintenance. At the Indiana University School of Medicine, Lawrence Wheat, Richard Kohler, and Ram Tewari (1986) had developed a radioimmunoassay for *H. capsulatum* var. *capsulatum* polysaccharide antigen, which became a useful new method for the rapid diagnosis of disseminated histoplasmosis. By 1991, Wheat et al. had established that two units or more of antigen levels strongly suggested histoplasmosis relapse. Wheat and the members of the NIAID-MSG (1993) assessed the efficacy and safety of 400 mg/day oral itraconazole in preventing histoplasmosis relapse after the successful induction of amphotericin B therapy in 42 AIDS patients. Results from this clinical trial suggested that "itraconazole, 200 mg twice daily, is safe and effective in preventing relapse of disseminated histoplasmosis in AIDS patients" (p. 610). The clearance of antigen correlated with clinical efficacy. At Washington University by using genetic markers, Eric Spitzer et al. (1990) found that AIDS patients were infected with less virulent strains (for mice) of *H. capsulatum* var. *capsulatum* in the St. Louis area. The efficacy of fluconazole in the treatment of histoplasmosis capsulati was investigated and it was concluded that 400 mg doses or higher were necessary for a successful outcome (McKinsey et al., 1996).

Coccidioidomycosis

At the University of Texas Health Sciences Center, San Antonio, John Graybill (1988), in collaboration with members of three other universities, led an evaluation of escalating high doses of ketoconazole (for up to four years) for the treatment of 15 patients with coccidioidal meningitis. It was demonstrated that high-dose ketoconazole alone or in combination with intrathecal amphotericin B had similar therapeutic value. However, a dose of ketoconazole higher than 800 mg/day was associated with intolerable nausea and vomiting. In 1990, the NIAID-MSG (Graybill et al.) elected to evaluate itraconazole as an alternative to the more toxic amphotericin B and ketoconazole for coccidioidomycosis. Oral itraconazole doses of 100–400 mg/day for up to 39 months were evaluated in the treatment of 49 patients with nonmeningeal coccidioidomycosis. An important accomplishment of this study was that the clinical response to therapy was evaluated with a scoring system accounting for lesion number and size, symptoms, culture, and serologic titer. A scoring system was needed to standardize criteria to measure clinical response; prior results had been based on variable criteria for defining complete remission and lesser degrees of improvement. Preliminary results indicated that itraconazole appeared to be superior to ketoconazole for the treatment of nonmeningeal coccidioidomycosis (Graybill et al., 1990).

Ketoconazole and miconazole became the second-line therapies for coccidioidal meningitis, the most fatal type of infection with *C. immitis*. The established therapy before the introduction of these agents was frequent administration of intrathecal amphotericin B. It was evident to John Galgiani and the NIAID-MSG members by 1990 that safer and more effective therapies were needed; intrathecal administration of this agent was technically difficult, produced irritation and discomfort, was associated with other infections, and was not always effective. With the advent of fluconazole and itraconazole, Galgiani and the NIAID-MSG members (1993) investigated the safety and efficacy of 400 mg/day up to four years of fluconazole in 50 patients with active coccidioidal meningitis. Thirty-seven of 47 (78%) evaluable patients responded to treatment within four to eight months. They concluded that "fluconazole therapy is often effective in suppressing coccidioidal meningitis" (p. 28).

Aspergillosis, candidiasis and sporotrichosis

David Denning, Richard Tucker, Linda Hanson, and David Stevens (1989) evaluated an oral itraconazole regimen of 400 mg/day in 18 patients with invasive aspergillosis at Santa Clara Valley Medical Center, CA. Their conclusion was that "itraconazole may be an important advance in the therapy of aspergillosis" (p. 791). Denning and the NIAID-MSG members in 1994 conducted a multicenter study to further evaluate itraconazole for the treatment of invasive aspergillosis. Seventy-six evaluable patients received oral itraconazole for 0.3 to 97 weeks. Their data showed that the response rate to oral itraconazole was comparable to that of amphotericin B therapy. The results suggested that oral itraconazole was an alternative therapy for this fungal infection, but relapse was noted among immunocompromised patients.

Patients with profound neutropenia and corticosteroid induced phagocyte dysfunction are susceptible to candidiasis and aspergillosis. At the NCI between 1990 and 1993, animal models were developed to evaluate the treatment of these infections in this patient population. Thomas Walsh et al. (1990) evaluated oral fluconazole for the prevention and early treatment of disseminated candidiasis in granulocytopenic rabbits. These investigators reported that fluconazole was as effective as amphotericin B for prevention, but not for treatment. Before 1993, desoxycholate amphotericin B was the treatment of choice for pulmonary aspergillosis in granulocytopenic patients. To reproduce the persistent levels of profound granulocytopenia in humans, Peter Francis et al. developed a model of primary pulmonary aspergillosis in rabbits in 1994. Utilizing this model, they evaluated the efficacy of a new formulation of unilamellar liposomal amphotericin B (approved in western Europe) and the detection of *Aspergillus fumigatus* D-mannitol in bronchoalveolar lavage fluid and galactomannan in serum. Both diagnostic approaches were useful markers of pulmonary aspergillosis. Since the new therapeutic agent was more effective and safer than desoxycholate amphotericin B, their results appeared to warrant the evaluation of these novel approaches in humans.

Evaluation of 100–600 mg/day itraconazole dosages for the treatment of 27 patients with sporotrichosis showed that this drug was effective in the treatment of cutaneous and systemic sporotrichosis (Sharkey-Mathis et al., 1993).

Antifungal susceptibility testing

Despite controversial views regarding the need and "right time" to have a subcommittee to set standards for antifungal susceptibility testing, the National Committee of Clinical Laboratory Standards (NCCLS) approved the formation of a subcommittee in April 1982 (NCCLS archival materials). John Galgiani of the University of Arizona became the chair of the subcommittee and in July, 1982, he proposed that the subcommittee be composed of six voting members and five advisors (J. Galgiani, personal communication, 1994). The first meeting of the NCCLS Subcommittee was held on Thursday, October 7, 1982 (NCCLS Subcommittee minutes, October 21, 1982). As a result of collaborative studies (Pfaller et al., 1990; Espinel-Ingroff et al., 1992; Fromtling et al., 1993), the subcommittee proposed a reference method for antifungal suscept-ibility testing of yeasts (NCCLS M27-P document) in December 1992, which became the approved standard in 1997 (NCCLS M27-A document). By 1994, the activities of the subcommittee had diversi-fied and four working groups were formed within the subcommittee. In 1996, the NCCLS subcommittee established tentative breakpoints for fluconacole, itraconazole (*Candida* spp.), and flucytosine based primarily, for fluconazole and itraconazole, on clin-ical data from oropharyngeal infections provided by the manufacturers of these two drugs (Rex et al., 1997). Its efforts were directed later toward the development of standard guidelines for the filamen-tous fungi and evaluations of more convenient procedures for routine use in the clinical laboratory. The subcommittee continues to investigate the uti-lity of reference methods for yeasts and moulds.

Laboratory diagnostic methods

During the 1980s and 1990s, The search for more sensitive, reliable, and rapid tests for the diagnosis of early invasive candidiasis and other fungal diseases continued. The following approaches were investi-gated for the diagnosis of early disseminated candi-diasis: detection of protein or cell wall mannoprotein marker antigens, the detection of the fungal metabo-lites D-arabinitol and enolase, and DNA probes. For this purpose, different methodologies were explored: solid-phase sandwich radioimmunoassay for the detection of soluble cytoplasmic protein antigens (P. Stevens, Huang, Young, and Ber-dischewsky, 1980), gas–liquid chromatography for the detection of mannose (Monson and K. Wilk-inson, 1981), latex coated particles with heteroge-neous antibodies (Gentry, Wilkinson, Lea, and Price, 1983; Ness, Vaughan, and Woods, 1989), combined microbiological and gas chromatographic methods to detect D-arabinitol (Bernard, Wong, and Armstrong, 1985), a sandwich enzyme immunoas-say to detect mannan (de Repentigny et al., 1985), a

combined gas chromatographic and enzymatic assay to detect D-arabinitol (Wong and Brauer, 1988), mouse monoclonal IgG to detect an immu-nodominant 48 kDa enolase antigen of *C. albicans* (A.B. Mason, Brandt, and Buckley, 1989), and a double-sandwich liposomal immunoassay for *Can-dida* enolase (Walsh et al., 1991).

Much progress has been accomplished since T.E. Kiehn et al. discovered arabinitol in the serum of patients with disseminated candidiasis in 1979 while they were testing for the presence of serum mannose. At the University of Cincinnati, Brian Wong and his co-workers have studied arabinitol as a possible diagnostic marker since the late 1980s. In 1988, Wong and Karen Brauer reported that infected patients had elevated serum levels of D-arabinitol and normal levels of L-arabinitol, thereby identify-ing the diagnostic marker. Wong, Jeffrey Murray, Miguel Castellanos, and Kenneth Croen (1993) clarified the D-arabinitol biosynthetic pentose path-way and cloned the gene that encoded the regulation of intracellular NAD-dependent D-arabinitol dehy-drogenase.

The following commercial kits became available to diagnose systemic candidiasis. From Baylor Col-lege of Medicine, Houston, (TX), the Cand-TEC latex agglutination system (Ramco Laboratories, Houston, Texas) was based on the detection of antigen as described by Gentry et al. (1983). The ICON *Candida* assay (ICON Hybritech Inc., San Diego, California) is an enzyme immunoassay. Even though the ICON appeared to be more reliable than the Cand-TEC, both kits were evaluated and found to be lacking in specificity and sensitivity (Burnie and Williams, 1985; Ness et al., 1989; Pfaller et al., 1993). In Japan, a commercial rapid test was devel-oped in the early 1990s for the enzymatic-fluoro-metric determination of serum D-arabinitol. An automated enzymatic method was described by Arthur Switchenko et al. (1994) for this purpose (Syva Company, Palo Alto, CA, National Cancer Institute, and University of Cincinnati). The reaction steps of the assay were adapted so that they could be performed automatically in the clinical laboratory on a chemistry analyzer. In this way, many of the problems associated with the current procedures for the detection of D-arabinitol, including the cross-reactivity associated with the enzymatic-fluoro-metric method, were avoided. Preliminary results demonstrated that detection of D-arabinitol per-mitted detection of invasive candidiasis and early recognition of fungemia in serially collected sera from high-risk cancer patients as well as therapeutic monitoring of the infection (Walsh et al., 1995).

At Audie Murphy Memorial Veterans Hospital in San Antonio, Marc Weiner (1983) developed a promising radioimmunoassay for the detection of *C. immitis* antigen. This assay detected antigenemia in five of nine patients (56%) with active coccidioidomycosis. These tests need further evaluation and may become valuable diagnostic and prognostic tests. They may complement rather than replace blood cultures.

George Hageage and Brian Harrington (1984) observed that calcofluor white, a non-specific fluorochrome with an affinity for chitin and cellulose, could be used to detect fungal elements in clinical specimens. This procedure eliminated some inaccuracies associated with KOH preparations as well as being faster than the selective PAS and silver stains. It requires a fluorescence microscope and it has replaced India ink and KOH preparations for the detection of fungi in the clinical microbiology laboratory.

DNA-based methods

At Montana State University, Jim Cutler, Pati Glee, and Harold Horn (1988) were the first to apply a DNA probe for the direct detection of *C. albicans* in clinical specimens. They isolated a DNA fragment that hybridized specifically with DNA from *C. albicans* but did not hibridize with DNA from other infectious agents or from the host. Another avenue that had been explored was the amplification of a DNA segment (probe) that lies within a fungus-specific gene by the polymerase chain reaction assay (PCR). Timothy Buchman, Michelle Rossier, William Merz, and Patricia Charache (1990), at the Johns Hopkins Hospital, reported the "unambiguous detection of *C. albicans* by PCR within 6 hours from the time that a clinical specimen is obtained" (p. 339). Roy Hopfer, P. Walden, S. Setterquist, and W. Highsmith (1993) simplified the sample preparation for the detection of 18 species of fungi, including *C. albicans*, from clinical specimens except blood. However, for these tests to be more convenient methods for the clinical laboratory, further simplification should be attempted. Also, their sensitivity in clinical specimen remains to be fully evaluated

Although detection of the dimorphic fungi in tissue and their identification by the exoantigen test were helpful for many years, these techniques were time consuming. The DNA probe assay for the identification of bacteria in the clinical laboratory was introduced during the mid-1980s. DNA probes were first adapted for fungal identification at the University of Tennessee for *Candida* spp. (including *C. glabrata*) (M.M. Mason, Lasker, and Riggsby, 1987) and at Washington University for *H. capsulatum* var. *capsulatum* (Keath et al., 1989). Several yeast form-specific genes were cloned and characterized by the latter investigators; the gene that may be associated with pathogenicity was their diagnostic probe. Another novel approach was the adaptation of the random amplified polymorphic (RAPD) DNA assay (Williams et al., 1990) for the characterization of isolates of *Candida* spp. by Paul Lehmann, Diming Lin, and Brent Lasker (1992) of the Medical College of Ohio and the CDC. They believed that genotype directed identification of yeasts could be the tool of the future (P. Lehmann, personal communication, April 10, 1994).

At Gen-Probe Inc. (San Diego, CA) Kathleen Clark developed a non-isotopic chemiluminescent probe system that simplified the early probe assays for the identification of fungi (K. Clark, personal communication, July 1994). This probe utilizes a single-stranded DNA probe that complements the ribosomal RNA of the fungus being identified. Commercial probes had been evaluated for the identification of *B. dermatitidis*, *C. immitis*, *C. neoformans*, and *H. capsulatum* var. *capsulatum* (Hall, Pratt-Rippin, and Washington, 1992; Stockman, Clark, Hunt, and Roberts, 1993).

DNA-based epidemiology

By 1987, various methods had been developed as epidemiologic tools to identify a specific strain as the source of infection. This early methodology relied on cumbersome, insensitive, and non-reproducible tests (streak morphology, resistance to antifungal drugs, enzymatic profiles, biochemical assays, etc.) that measured phenotypic characteristics. DNA-based typing procedures, such as molecular probes or the determination of electrophoretic karyotypes, replaced earlier methodologies. The development of pulsed-field gel electrophoresis procedures demonstrated that the chromosome mobility of *C. albicans* and *S. cerevisiae* was extremely variable under apparently identical testing conditions. At the CDC, Timothy Lott, Patrick Boiron, and Errol Reiss

(1987) defined an electrophoretic karyotype for *C. albicans* by field-inversion gel electrophoresis, while Lott was conducting his postdoctoral studies with Reiss. This karyotype, the result of the migration of intact chromosomes, was distinct from other species of *Candida* and thus species-specific. Beatrice Magee and Paul T. Magee (1987) at Michigan State University used orthogonal field alternating gel electrophoresis to demonstrate that isolates of *C. albicans* and *Candida* non-*albicans* had different electrophoretic karyotypes. In 1989, C.S. Kaufmann and William Merz at the Johns Hopkins Hospital defined different electrophoretic karyotypes among isolates of *C. glabrata*. In the same year, John Perfect, B.B. Magee, and P.T. Magee (1989) identified different electrophoretic karyotypes of *C. neoformans*, while Perfect had his one year sabbatical with P.T. Magee at the University of Minnesota (J. Perfect, questionnaire; "Training Trees", Appendix B). These studies suggested that electrophoretic karyotypes could be used as "stable genetic markers to delineate strains for use in epidemiologic studies" (C.S. Kaufmann and Merz, 1989, p. 2165).

Recombinant DNA technology also made possible the replacement of phenotyping methodology by the direct examination of the genomes of the fungi (Tompkins, Plorde, and Falkow, 1980). High resolution for this procedure was obtained by molecular probes. Stewart Scherer and David Stevens (1987), at the University of Minnesota and Santa Clara Valley Medical Center, used this approach to study *C. albicans* and described a DNA extraction and digestion method for the whole *Candida* genome. By using restriction endonuclease digestion and electrophoresis, "chromosomal DNA of various restriction lengths in gel produced unique patterns analogous to fingerprints" (p. 675). These gel patterns could be used to type *Candida* spp. isolates, which was a more precise epidemiologic tool than the ones based on analysis of phenotypic characteristics. By 1988, Scherer and Stevens reported that with their 1987 procedure, "only a limited number of sites in the genome of a given species can be scored for differences" (p. 1452). Therefore, they isolated a repeated DNA segment that provided a species-specific DNA probe for *C. albicans*. At the University Hospital, Stony Brook (NY), Eric D. Spitzer and Silvia G. Spitzer (1992) demonstrated that the CNRE-1 hybridization probe could discriminate among 10 strains of this species isolated from eight patients. Based on repetitive DNA sequences from *C. neofor-*

mans, this probe could demonstrate that the cryptococcal isolate from the original infection and relapsing isolates were identical. Thus, the use of DNA probes as another epidemiologic tool allowed measurement of DNA polymorphism in the laboratory. Controversy began regarding which methodology was more sensitive and the CHEF appeared to provide greater sensitivity than the enzymatic tests in recognizing strain variation (Vazquez et al., 1993).

In addition, DNA fingerprinting and random amplified polymorphic DNA (RAPD or PCR) were adapted as epidemiologic tools. The RAPD was described by John Williams, Anne Kubelik, Kenneth Livak, J. Antoni Rafalski, and Scott Tingey (1990) at the Central Research and Development Department, E.I. du Pont de Nemours and Co., Inc., Wilmington, DE. Since DNA purification for the RAPD assay was time consuming, Jon Woods, Dangeruta Kersulyte, William Goldman, and Douglas Berg (1993) at Washington University shortened their earlier methodology (1992) for *H. capsulatum* var. *capsulatum* DNA purification from a 2-day to a 2-h procedure. Investigators at Duke University, Wieland Meyer, Thomas Mitchell, Elizabeth Freedman, and Rytas Vilgalys (1993), combined RAPD and conventional DNA fingerprinting to detect *C. neoformans* strain variation during Meyer's postdoctoral training with Mitchell ("Training Trees", Appendix B). They found that, for detecting polymorphic DNA in *C. neoformans*, this combination yielded more reliable results than "classical" DNA procedures alone. In the same year, Timothy Lott, Randall Kuykendall, Sharon Welbel, Arum Pramanik, and Brent Lasker at the CDC (1993) demonstrated that pulsed-field gel electrophoresis and RAPD can be used to show *C. parapsilosis* intraspecies variation. Brian Currie, Lawrence Freundlich, and Arturo Casadevall (1994) used fragment length polymorphism analysis with two probes to characterize and compare *C. neoformans* isolates from pigeon excreta with those isolated from human cryptococcal meningitis (same geographical area). Although they found extensive genetic diversity among environmental and clinical isolates, they could not answer the question of whether pigeon excreta are a reservoir for yeasts causing human infection.

Taxonomy and classification

Michael McGinnis, at the University of North Carolina, stated in 1983 that the "nomenclature for the mycotic infections caused by the black fungi is in a state of confusion" (p. 1). He concluded that clarification was needed regarding the names chosen for the infections known as chromoblastomycosis and phaeohyphomycosis. McGinnis believed that these fungal infections should be named based upon the combined clinical, pathologic, and mycologic relationships of each infection. McGinnis (1983) defined chromoblastomycosis as a chronic, localized cutaneous or subcutaneous infection. The infected tissues contain muriform cells and histologically show pseudoepitheliomatous hyperplasia, dermal granulomas, abscess formation, and fibrosis. The term phaeohyphomycosis had been described by Libero Ajello et al. (1974) to include mycotic diseases in which the etiologic agent (heterogeneous group) was present in tissue as phaeoid yeast, pseudohyphae or hyphae (and no muriform cells). This term was not to include nor replace mycetoma pityriasis (tinea nigra), black piedra or mycotic keratitis. In 1982, McGinnis, Arvind Padhye, and Ajello established that *Pseudallescheria boydii* was the valid name for this fungus.

Other taxonomic problems were clarified in 1986. McGinnis, Dante Borelli, Arvind Padhye, and Ajello (1986) proposed that the hyphomycete *Cladosporium bantianum* be transferred to the genus *Xylohypha* as *Xylohypha bantiana* based on morphological characteristics. In 1995, G.S. de Hoog et al. transferred this species to the genus *Cladophialophora*. McGinnis, Rinaldi, and Winn (1986) studied isolates of the genera, *Bipolaris*, *Drechslera*, *Exserohilum*, and *Helminthosporium* and clarified the taxonomy surrounding human pathogenic species belonging to those genera. Since the early 1970s, there had been an increase in the frequency of infections caused by these fungi, which were easily misidentified in the clinical laboratory. Excellent descriptions of diagnostic, morphologic features, and photographs of each species were included in this report. Irene Weitzman, McGinnis, Padhye, and Ajello (1986) concluded that the genera *Arthroderma* and *Nannizzia* were congeneric. Because of priority, *Nannizzia*, which was established by Phyllis Stockdale in 1963, was considered a synonym of *Arthroderma* (proposed by Currey in 1854).

Vaccines

Beginning in the late 1950s, vaccines were made to immunize guinea pigs and human volunteers against *Trichophyton mentagrophytes* (Huppert and Keeney, 1959) and monkeys against *C. immitis* (Pappagianis, Miller, C.E. Smith, and Kobayashi, 1960; Naval Biological Laboratory, California). Both studies showed that these primitive vaccines could stimulate an immune response. Thirty years later, a study of 2867 healthy human volunteers randomized to either a vaccine (*C. immitis)* group (1436 subjects) or a placebo control group (1431 subjects) showed that the differences between the two groups were not statistically significant. This study was conducted between 1980 and 1985 by the Valley Fever Vaccine Study Group and was published in 1993. Demosthenes Pappagianis et al. (1993) believed that the fungal fractions (e.g., proteinases) isolated from *C. immitis* in the 1980s and 1990s could produce a more effective vaccine in the future.

Host defenses against yeasts

Cryptococcus neoformans

Between 1980 and 1993, important findings were reported regarding the host-parasite interaction variables in cryptococcosis by Juneann Murphy and her graduate and postdoctoral students at the University of Oklahoma at Norman and Oklahoma City ("Training Trees", Appendix B). Murphy focused her research on the host defenses against *C neoformans*, the lymphocytes, especially CD4[+] T lymphocytes involved in the cell-mediated immune response specific to *C. neoformans*, and phagocytic and natural killer cells. This group reported for the first time the developmental profiles of the delayed-type hypersensitivity (DTH) response (a measurement of cell mediated immunity), antigen levels, and antibody titers induced after intranasal inoculation of mice with *C. neoformans* (Lim, Murphy, and Cauley, 1980). The investigation of cell-mediated response had begun in the late 1970s, while Murphy was conducting research for her doctoral dissertation under the guidance of George Cozad ("Training Trees", Appendix B).

Thuang Lim et al. (1980) demonstrated direct correlations between high *C. neoformans* antigen levels and a depressed DTH response. It was reported by Mosley, Murphy, and Cox (1986) that intravenous injection of *C. neoformans* antigen triggered the induction of a cascade of *C. neoformans* suppressor cells (T cells) and factors that downregulated the protective anticryptococcal cell-mediated immune response. Of this "cascade" of suppressor T cells (Ts1, Ts2, and Ts3), Ts3 worked in conjunction with Ts2 to inhibit the anticryptococcal DTH response in the murine model. The Ts1 cells induced the Ts2 cells and so on (Khakpour and Murphy, 1987). It was Murphy's hypothesis that in humans, who have high serum levels of *C. neoformans* antigen, a similar induction of suppressor cells might occur. In 1988, Murphy and investigators at the CDC, Georgia State and at the University of Nevada reported that mannoprotein was the primary component recognized by the anticryptococcal cell-mediated immune response in mice. In 1989, Murphy and her student Paul Fidel found that cyclosporin A, a potent immunosuppressive drug, affected DTH cells (suppression effect).

Following Juneann Murphy's demonstration that cell-mediated immunity is a key host cell defense in cryptococcosis, her research focused on defining the lymphokines associated with induction and regulation of the anticryptococcal cell-mediated immune response. Her studies have shown that in mice immunized with cryptococcal antigen, two other populations of CD4$^+$ T cells were induced in addition to suppressor cells: ^{T}DH cells, responsible for the anticryptococcal DTH response, and Tamp cells, responsible for amplification of this response. Murphy's data (1993) showed that "spleen cell populations that contain ^{T}DH and Tamp cells produced more lymphokines (gamma interferon and interleukin-2 [IL-2]) than spleen cell populations that contained only ^{T}DH cells" (p. 4750). These lymphokines were influential in the development of the anticryptococcal response amplified by the Tamp cells. Murphy (1993) stated that these results may serve as a basis for future studies in understanding the mechanisms by which lymphokines affect clearance of *C. neoformans* from infected human tissues.

The activity of the cytotoxic effector cells, or natural killer (NK) cells, had been described since the 1970s by R. Kiessling, E. Klein, and H. Wigzell (1975). NK cells were found in lymphoid tissues, but were absent at birth and in the thymus of normal, unimmunized individuals. The role of NK cells as primary mammalian host defense mechanisms against viral and tumor targets had been studied,

but their role as nonphagocytic killer cells, a natural defense mechanism in infectious diseases, had not been established by the early 1980s.

In 1982, substantial but indirect evidence was provided by Murphy and Olga McDaniel that unstimulated murine splenic cells with the characteristics of NK cells could inhibit the in vitro growth of *C. neoformans*. Therefore, NK cells were potentially the third means of natural cellular immunity, the other two being macrophages and neutrophils. By 1985, Murphy's group had "elucidated" some mechanisms by which NK cells affected *C. neoformans*. In vitro, NK cells bound and formed conjugates with *C. neoformans*, with the number of these conjugates being directly proportional to the degree of growth inhibition (Nabavi and Murphy, 1985). In 1991, Michelle Hidore, Nasrin Nabavi, F. Sonleitner, and J.W. Murphy demonstrated the participation of NK cells in the clearance of *C. neoformans* in vivo by exocytosis of cytolytic material. This process resulted in the death of *C. neoformans* and was the same sequence of events observed in the interactions of NK cells with tumor targets. Murphy, Hidore, and Si Chai Wong had shown by 1993 that human NK cells, monocytes, and T lymphocytes bound and inhibited the growth of *C. neoformans* cells in the absence of cryptococcal antibody or complement (a nonphagocytosis process).

Murphy's group also became interested in the potential surface components of *C. neoformans* involved in chemotaxis of neutrophils. Zhao Dong and Murphy (1993) had hypothesized that two chemotactic factors were involved; one was *C. neoformans* derived, and the second was serum derived. Both factors were activated by cryptococcal components. It was then concluded that capsular glucuronoxylomannan was a contributor to the direct chemotactic activity of human neutrophils and that capsular mannoprotein only stimulated indirect chemotatic activity. At the University of Nevada, Thomas Kozel, M.A. Wilson, G.S.T. Pfrommer, and A.M. Schlageter (1989) demonstrated that the classical complement pathway had little or no role in the opsonization of *C. neoformans*.

Candida albicans

At the University of South Florida, Julie Djeu directed her research toward the antifungal activity of neutrophils against *C. albicans*. She investigated the factors released by T lymphocytes and natural killer cells (as large granular lymphocytes) that induced killing by neutrophils. By using a rapid radiolabel microassay, Djeu measured the growth inhibition of *C. albicans* by lymphokine regulation of neutrophils. By 1986, Djeu, Kay Blanchard, Demetrios Halkias, and Herman Friedman had shown that two lymphokines, tumor necrosis factor and gamma interferon, could activate neutrophils. They also reported that neutrophils were better than peripheral blood lymphocytes in inhibiting growth of *C. albicans* in vitro. In addition, natural killer cells could produce those two lymphokines and granulocyte-macrophage colony stimulating factor (GM-CSF), among others. These factors had the ability to directly activate and mobilize neutrophils. Since the neutrophil-activating activity in natural killer cells was not neutralized by antibodies to tumor necrosis factor and gamma interferon, another factor was also involved. By 1991, Blanchard, Beatriz Michelini-Norris, and Djeu had identified natural killer cells "as the source of neutrophil-activating factors, of which GM-CSF played a central role as a mediator between natural killer cells and neutrophils" (p. 2259). This factor was not produced by monocytes or small mature T cells. Djeu noted the possibilitiy that these cells could produce other cytokines following activation by *C. albicans* and that further research was warranted to better understand the role of neutrophils in host resistance to infection.

In 1988, Lehrer, Ganz, Szklarek, and Selsted reported that human neutrophils contained four defensins (six in rabbits), and that only two of them, HNP-1 and HNP-2, killed *C. albicans* effectively in vitro.

Meanwhile, Richard Calderone, at Georgetown University, focused his research on the identification of *C. albicans* receptors that bind complement as well as the potential pathogenic role of these receptors. Several membrane glycoproteins (CR1, CR2, CR3, and CR4) are found on human peripheral blood cells. These glycoproteins serve as receptors for the binding of cleavage products of the third (C3) component of complement (e.g., CR2 binds C3d) and fungal cells also have similar specific receptors. In 1988, Calderone, Lisa Linehan, Elsa Wadsworth, and Ann Sandberg initiated for the first time the investigation of the functional significance of the receptors for C3 complement fragments on *C. albicans*. They identified two protein receptors, IC3b (7 kDa) and C3d (60 kDa), that bound the corresponding C3 fragments of complement in *C. albicans* hyphal extracts. It was shown in 1991 by immunofluorescence and immunoelectron microscopy technology, that the C3d receptors (CR2) were produced by both forms of *C. albicans* in vitro and in vivo (murine model) conditions (Toshio Kanbe, Ren-Kai Li, Elsa Wadsworth, Calderone, and Jim Cutler from Montana State University and Georgetown University). The receptors of blastoconidia were buried at the level of the plasma membrane. By 1993, Calderone's group had purified the receptors, mannoproteins, from blastoconidia (50 kDa) and hyphae (60 kDa) of *C. albicans*. Both mannoproteins inhibited the binding of antibody-sensitized sheep erythrocytes conjugated with iC3b or C3d by hyphae of *C. albicans*. Wadsworth, Prasad, and Calderone reported in 1993 that although the two mannoproteins had dissimilar properties, they had a common antigenic determinant.

At Tulane University, the research focus of Judith Domer and her students was to examine the regulation of the cellular immune responses mediated by candidal cell wall mannan and glycoproteins in a murine model. In 1989, Domer, R.E. Garner, and R.N. Befidi-Mengue extracted mannan from *C. albicans* and found that the cell-mediated response to mannan was greater than the response to cell wall glycoprotein in vivo. The effect was reversed in vitro. However, it was unclear which component of the mannan extract, e.g., mannan, protein, or the intact mannoprotein, "was responsible for the detection as well as the suppression of cell-mediated immunity" (Domer et al., 1989, p. 697). Therefore, their conclusion was that complete separation of the mannan component from the protein component were essential to clarify this issue.

Oropharyngeal and gastric infections with *Candida* spp. occur frequently in AIDS patients, however, in the early stages of AIDS systemic candidiasis is rare. Because of that, development of an animal model of a retrovirus-induced immunodeficiency syndrome was developed to examine the exacerbation of gastric candidiasis. Garry Cole and K. Saha et al. (1992) in collaboration with the M.D. Anderson Cancer Center and the University of Sao Paulo (Brazil) reported the first murine model of invasive gastrointestinal candidiasis associated with an AIDS-related murine immunodeficiency syndrome. Infant mice were infected by oral-gastric inoculation with the retrovirus complex. This model would allow investigations of *Candida* infections in the immuno-suppressed host during progressive stages of AIDS-associated induced infection in mice.

Host defenses against the dimorphic fungi

Coccidioides immitis

It had been shown by Demosthenes Pappagianis and coinvestigators (Beaman, Pappagianis and Benjamini, 1977) at the University of California, Davis that T lymphocytes were essential in protecting mice against *C. immitis*. Although these cells did not kill the fungus, they had a role in initiating its killing by macrophages (phagosome-lysosome fusion). By 1983, LoVelle Beaman, E. Benjamini, and Pappagianis demonstrated that the release of lymphokines was induced by exposure of macrophages to antigen-stimulated splenic lymphocytes. This killing mechanism, which had not been previously described, activated macrophages thus enhancing phagocytosis and killing.

In 1991, Theo Kirkland et al. reported that a soluble conidial wall fraction (SCWF) of *C. immitis* stimulated murine T cells in vitro. Their generation of an antigen-specific murine T-cell line for SCWF demonstrated that the most antigenic subfractions of SCWF had molecular masses of 43 to 66 kDa. A portion of the gene, which encoded one T-cell antigen, was cloned and expressed as a lambda gt11 fusion protein by Kirkland et al. (1991). Previously, three clinically relevant antigens (diagnostic serology) from coccidioidin (the immunodiffusion and precipitin antigen [IDTP], the immunodiffusion and complement-fixing antigen [IDCF], and the heat-stable antigen in extracts of mycelium [IDHS] had been isolated and identified by Rebecca Cox and Lorene Britt (1987) at the San Antonio State Chest Hospital. By using two-dimensional immunoelectrophoresis, Cox and Britt (1987) demonstrated that these three antigens were contained in both spherulin and coccidioidin preparations.

Histoplasma capsulatum var. *capsulatum*

At the University of California, Los Angeles, Dexter Howard and his graduate and postdoctoral students ("Training Trees", Appendix B) continued investigating the host defense mechanisms in histoplasmosis capsulati. By 1984, Betty Wu-Hsieh and Howard had observed that lymphokines armed macrophages against *H. capsulatum* var. *capsulatum* in vitro and that lymphokines had a high interferon activity (heat stable and acid labile). In 1994, Nakamura, Wu-Hsieh, and Howard demonstrated that recombinant murine gamma interferon "was the key element within the lymphokine-containing supernatants which was responsible for activating macrophages to the growth inhibitory state" (p. 683). This activation required the metabolism of arginine. George Deepe focused his investigations on the immune disturbances associated with *H. capsulatum* var. *capsulatum*. He conducted these studies under Ward Bullock at the University of Cincinnati, where he did his postdoctoral training (Dalldorf Fellow 1981–1983) ("Training Trees", Appendix B). Deepe first isolated and propagated *H. capsulatum*-reactive murine T-cell lines (TCL, CD3[+] and CD4[+]) and clones. These cells released IL-2 and a factor that stimulated macrophages to limit *H. capsulatum* growth in vitro. Deepe demonstrated in 1988 that neither TCL nor cloned T cells could transfer resistance to this fungus nor confer a DTH response in vivo. However, CD4[+] cells from immunized mice could enhance an immune response.

Blastomyces dermatiditis

At the Santa Clara Valley Medical Center, Elmer Brummer's focus of investigation was the phagocytic host defense mechanisms of resistance in blastomycosis. David Drutz and C.L. Frey (1985) had previously reported the importance of human phagocytes as a nonspecific defense against *B. dermatitidis* conidia and hyphae; however, the parasitic form (yeast) was resistant to killing by phagocytes. By 1984, Brummer and David Stevens had hypothesized that neutrophil fungicidal activity was specifically enhanced by T-cell products, such as chemotactic factors (lymphokines), and by leukocyte and monocyte migration-inhibitory factors. Their investigation demonstrated for the first time a link between the soluble factors produced by stimulation of sensitized lymphoid cells with *B. dermatitidis* antigen and the induction of enhanced antimicrobial activity of neutrophils against *B. dermatitidis* in vitro. In 1992, Brummer and his colleagues at Chiba University (Japan) investigated several mechanisms of resistance of *B. dermatitidis* to killing by neutrophils. Their data suggested that resistance to killing was dependent on the "inefficient generation of products from the peroxidase-dependent neutrophil microbicidal system" (Brummer, N. Kurita, S. Yoshida, K. Nishimura, and M. Miyaji, 1992, p. 233). Brummer had previously shown in the mid-1980s that lymphokines could modulate host phagocyte fungicidal activity in vitro and that purified interferon gamma stimulated the fungicidal activity of macrophages and neutrophils.

The in vitro interaction between macrophages and their activities and azole antifungal agents against *B. dermatitidis* was investigated in 1992 by Brummer, Purushothama Bhagavathula, Linda Hanson, and Stevens. They demonstrated that itraconazole fungistatic concentrations in combination with murine peritoneal macrophages synergistically killed *B. dermatitidis*. In 1993, Brummer, Hanson, and Stevens also reported the direct relationship between disease progression (blastomycosis) and elevated serum levels of IgE with the production of the IL-4 lymphokine by antigen-stimulated murine spleen cells. In contrast, they observed an inverse relationship between serum IgE and gamma interferon levels, that was associated with resistance to infection and healing. Elevated IgE serum levels had been reported in association with chronic parasitic disease progression in mice. Brummer et al. (1993) extended for the first time this association to murine blastomycosis. A new animal model for the study of immunoregulation in fungal diseases was thus postulated by inducing a chronic infection in immunized mice. At Boston University Medical Center, Alan Sugar and Michele Picard (1991) used Brummer's animal model that was developed in the early 1980s to demonstrate that at 37°C both macrophages and H_2O_2 could selectively block transition of germinating conidia to *B. dermatitidis* yeast form.

Host defenses against Aspergillus fumigatus and Rhizopus arrhizus

In human and experimental animal infections, neutropenia and neutrophil or macrophage dysfunction are the greatest risk factors to invasive aspergillosis. As a Dalldorf Fellow (1984–1986), Stuart Levitz focused his research on the role of the phagocytic host defense (neutrophils and macrophages) in aspergillosis and zygomycosis. Levitz conducted his postdoctoral studies under Richard Diamond's tutelage at Boston University ("Training Trees", Appendix B). In 1985, Levitz and Diamond reported the extreme resistance of the resting *A. fumigatus* con-

idial cells to oxidative and nonoxidative fungicidal products of neutrophils. They found that in addition to their natural resistance, a suboptimal release of neutrophil products was stimulated by resting conidial cells. The following year, Levitz, Michael Selsted, Thomas Ganz, Robert Lehrer, and Diamond demonstrated that the dormant, resting *A. fumigatus* and *Rhizopus arrhizus* (as *A. oryzae)* conidial cells were also not killed by rabbit neutrophil and defensins (macrophage antimicrobial cationic peptides) (Selsted, Brown, DeLange, Harwig, and Lehrer, 1985). However, the fungicidal activity of phagocytic cells and their cationic peptides was acitvated once the invasive form of these fungi began to grow. These results corroborated Diamond's group data reported in the late 1970s that neutrophils and macrophages attached to hyphae and caused damage by release of oxidative metabolites.

Waldorf, Levitz, and Diamond (1984) found that bronchoalveolar macrophages from normal mice participated in defense against *R. arrhizus* (as *R. oryzae*) by inhibiting the germination of conidia or conversion to the tissue-invasive stage. In contrast, defense against *A. fumigatus* was not dependent on inhibition of conidial germination but on early killing of conidia. In diabetic and cortisone-treated mice, bronchoalveolar macrophages allowed conidial germination or infection by *R. arrhizus*. In the cortisone-treated mice, macrophages did not kill fungal conidia, and aspergillosis infection occurred. Levitz et al. (1986) concluded, "it is only when phagocytic defenses are profoundly disturbed that invasive infections are seen" (p. 488).

HIV- (human immunodeficiency virus) infected patients are susceptible to invasive aspergillosis (without neutropenia or corticosteroid therapy risk factors) and neutrophils are an important component of the host defense in aspergillosis. This knowledge led Emmanuel Roilides, Andrew Holmes, Cassann Blake, Philip Pizzo, and Thomas Walsh (1993) to investigate the antifungal activity of neutrophils in 31 HIV-infected children. They also examined whether these patients' serum circulating factors and active HIV viral proteins were able to suppress neutrophil function. In addition, they investigated the possibility of improving the impaired patient's neutrophil antifungal function with the granulocyte colony-stimulating factor (G-CSF). It was then concluded by Roilides et al. (1993) that patients with low CD4 cell counts (<25% of normal median value) had impairment of serum-mediated antifungal activity; the defective function of neutrophils was partially corrected by G-CSF in five patients. In the same year, the hypothesis that corticosteroids had a deleterious effect on neutrophil antifungal function was evaluated by Roilides, Katrin Ühlig, David Venzon, Pizzo, and Walsh (1993). They investigated their hypothesis by studying the potential preventive utility of G-CSF and gamma interferon and concluded that "corticosteroids impair neutrophil function in response to *A. fumigatus* and that G-CSF and gamma interferon prevent this impairment" in vitro (Roilides et al., 1993, p. 4870). This mechanism has potential clinical significance, because it may serve as an adjunct to antifungal chemotherapy for the prevention and treatment of invasive aspergillosis in corticosteroid-treated patients.

Pathogenicity: Fungal enzymes

John Rippon's association of enzymes with virulence in the late 1960s led various investigators to investigate the role of different fungal enzymes in pathogenicity. At the University of Nevada, David Lupan and Pasipanodya Nziramasanga (1986) reported that *C. immitis* had collagenolytic and elastinolytic activity in vitro. At the University of Texas, Austin, Ling Yuan and Garry Cole (1987) characterized a single proteinase from *C. immitis* with those two activities. This enzyme can degrade human immunoglobulins and could "represent an important virulence factor in the development of coccidioidomycosis" (Yuan and Cole, 1987, p. 1970). Subsequent data developed by Cole and his graduate students ("Training Trees", Appendix B) suggested that the purified proteinase had a molecular size of 34-kDa and that the enzyme was localized in the cell walls of the *C. immitis* parasitic form (immunoelectron microscopic studies). Cole et al. isolated and expressed in 1992 the gene that encoded the enzyme. In the same year, David Kruse and Cole isolated and characterized the alpha-glucosidase enzyme that was "identical to the 120-kDa tube precipitin (TP) antibody-reactive mycelial fraction of *C. immitis*" (p. 4350). Both forms of *C. immitis* were able to express this enzyme. These facts suggested to Kruse and Cole (1992) that this enzyme could be responsible for plasticization of the cell wall that leads to spherule differentiation during the rapid diametric growth of the parasitic form.

Demosthenes Pappagianis and his graduate students ("Training Trees", Appendix B, University of California, San Francisco and Davis) also investigated the enzymology of *C. immitis*. Steven Resnick, Pappagianis, and James McKerrow (1987) isolated and characterized two major proteolytic components of *C. immitis*, a potent serine elastase and a metalloproteinase, and more importantly they hypothesized their potential association with virulence. A chitinase was isolated from *C. immitis* spherule-endospores by Suzanne Johnson, Charles Zimmerman, and Pappagianis (1993). Based on biochemical properties and serologic activity, this chitinase was identical to the complement fixing (CF) antigen. Amino-terminal protein sequence analysis of the chitinase linked the IDCF antigen with chitinase activity.

In the 1960s, Fritz Staib (1965) from Germany discovered a candidal proteinase. This important finding led to the characterization of candidal proteinase isoenzymes by R. Rüchel, also in Germany, in 1981; three strain dependent enzymes were found by Rüchel. In 1985, Kyung Joo Kwon-Chung, Donna Lehman, Carol Good, and Paul Magee used a murine model to examine the virulence of a proteinase-deficient mutant, its parent, and one proteinase-producing revertant. They concluded that the extracellular proteinase produced by *C. albicans* is one of the virulence factors associated with this organism" (Kwon-Chung et al., 1985, p. 571). There was a good correlation between the degree of virulence and the level of proteinase produced. At the CDC, Timothy Lott, L.S. Page, P. Boiron, J. Benson, and Errol Reiss (1989) reported the sequence of the *C. albicans* aspartyl proteinase (AP) gene; a total of 1705 bp were included in this proteinase. Based on evidence collected in England by Sullivan and Wright (1992), that more than one gene encoded the extracellular AP of *C. albicans*, Christine Morrison et al. (1993) examined the possibility that more than one protein was present in AP preparations. They detected three dominant proteins (41, 48, and 49 kDa) by sodium dodecyl sulfate-polyacrylamide gel electrophoresis in purified AP preparations. The CDC, the Century Children's Hospital, Mexico City and Emory University, Atlanta conducted this study.

The potential relationship between the extracellular acid proteinase from *C. albicans* and virulence led to the cloning of genes for this enzyme in four laboratories in Europe, New Zealand and the United

States. The gene that is regulated by switching (between opaque and white colony formation) in *C. albicans* codes an acid proteinase (Morrow et al., 1992). This gene is identical to the one isolated by B. Hube and his collaborators in Europe in 1991. These results indicated that different functions could be produced by two homologous genes. Beatrice Magee collaborated with colleagues in Europe and New Zealand (B. Magee, Hube, Wright, Sullivan, and P.T. Magee, 1993) and assigned the clones of the extracellular proteinase genes to the electrophoretic karyotypes of *C. albicans*. They also searched for two genes to be present. Their conclusion was that the "two genes map to different chromosomes, that every strain tested so far contains both genes, and that the nonidentity of the two genes was confirmed" (B. Magee et al., 1993, pp. 3240–3241). In addition, they found evidence for additional genes which suggested that there was a gene family in *C. albicans*. These investigators named this "gene family" the "secreted acid proteinase (SAP) family".

Molecular genetics

The genetics of *C. albicans* were extensively studied in the 1980s. Although auxotrophic mutants had been isolated and characterized since the 1970s, mapping and recombination studies were only initiated in the 1980s. The most influencial discovery was the recognition that *C. albicans* was a diploid organism when studied by UV-induced mitotic recombination that yielded auxotrophs (Whelan, Partridge, and P.T. Magee, 1980, at Michigan State University). This discovery corroborated an earlier conclusion by Ayodele Olaiya and Stephen Sogin (1979) that was only based on DNA content. It also led to significant progress in fungal genetic analysis. Another important contribution in this area of research was evidence from Wichita State University that the protoplasts of complementing auxotrophs of *C. albicans* could fuse in the presence of polyethylene glycol and generate prototrophic cells (Saracheck, Rhoads, and Schwarzhoff, 1981). Another tool for the study of fungi by genetic recombination had thus been developed.

During the late 1970s, DNA transformation and gene replacement in fungi with *S. cerevisiae* as a model began to appear in the literature. The yeast transformation system allowed the introduction of cloned DNA sequences into chromosomes by homologous recombination. This important prop-

erty led to the development of procedures for precise replacement of chromosomal regions with in vitro-altered DNA sequences (e.g., gene replacement). Because of that, Bruce Miller, Karen Miller, and William Timberlake (1985) demonstrated that *A. nidulans* genes can be replaced with mutant alleles made in vitro by either the one- or two-step procedure developed by Stewart Scherer and Ronald Davis (1979) for *S. cerevisiae*. Gene replacement permitted "rapid proof of the identity of new genes that are isolated by complementation of mutations" (B. Miller et al., 1985, p. 1714). It also made possible the investigation of the biochemical and biological consequences of introducing a specific, preselected mutation into the genome of otherwise unaltered cells.

At the Squibb Institute for Medical Research, Myra Kurtz, Mark Cortelyou, and Donald Kirsch (1986) developed a DNA-mediated transformation system that further improved the genetic analysis of *C. albicans* and other fungi. Because transformation was uncommon, "this system requires a gene that functions in the host and a system for selecting individuals that have taken up and expressed this gene" (p. 142). Kurtz et al. (1986) chose the cloned *C. albicans* ADE2 gene for transformation. This plasmid DNA became stably integrated into host DNA at the site of the ADE2 gene and became the first DNA-mediated transformation in *C. albicans*; it facilitated the creation of selective markers and was a valuable tool for the genetic manipulation of this yeast, and later led to the transformation of other fungi. Jeffrey Edman and Kyung Joo Kwon-Chung (1990) from the University of California, San Francisco and the NIH, respectively, isolated the marker URA5 gene from *C. neoformans*.

Research was directed towards cloning the genes of *C. albicans* when the transformation system of this yeast became available. Libraries of *Candida* DNA for complementation of auxotrophic mutations led to the identification of a significant number of DNA sequences. Simultaneously, pulsed field techniques of orthogonal field alternating gel electrophoresis (Schwartz and Cantor, 1984; Carle and Olson, 1984), field inversion gel electrophoresis (Carle, Frank, and Olson, 1986), and contour-clamped homogeneous field electrophoresis (CHEF) (Chu, Vollrath, and Davis, 1986) were developed. These procedures allowed the visualization of chromosomes. David Schwartz and Charles Cantor (1984) at Columbia University and C.F. Carle and M.V.

Olson (1984) developed pulsed-field gradient electrophoresis, a new agarose gel electrophoresis system capable of fractionating up to 2-Mb chromosome molecules. With this system, the chromosomes of *S. cerevisiae* were separated intact for the first time. By use of Southern blotting (Southern, 1975) and hybridization with chromosomally assigned probes, "gel bands can be assigned to their corresponding chromosomes" (Schwartz and Cantor, 1984, p. 67). Two years later at Stanford University, Gilbert Chu, Douglas Vollrath, and Ronald Davis (1986) manipulated electric fields by "arranging multiple electrodes along a polygonal contour and clamped to predetermined electric potentials" (p. 1582). The CHEF overcame the problem of a non-uniform electric field. DNA molecules larger than 2-Mb were resolved by Vollrath and Davis (1987) with the CHEF technique. Because pathogenic yeasts contain chromosomes that are larger than 2-Mb, the CHEF procedure became an important breakthrough.

It was suggested at the University of Minnesota and Smith Kline and French Laboratories (PA) by Beatrice B. Magee, Y. Koltin, J.A. Gorman, and Paul T. Magee (1988) that *C. albicans* had seven chromosomes; pulsed-field electrophoresis procedures allowed this conclusion. They also assigned genes to these chromosomes by hybridization. In the ensuing year, other identified karyotypes gave an estimate of six to nine chromosomes. Brian Wickes et al. (1991), at the NIH and the University of Minnesota, utilized more advanced pulsed-field electrophoresis methodology and studies of genetic linkages to report that the basic number of chromosomes for *C. albicans* was eight. Failure to separate the two largest chromosomes (including the rDNA gene) had led to a previous count of seven. During the same year, Wickes, J.E. Golin, and Kyung Joo Kwon-Chung (1991) demonstrated that phenotypic change occurred in conjunction with chromosomal rearrangement in *C. stellatoidea* type I (a subspecies of *C. albicans*). Kwon-Chung, Jeffrey Edman, and Wickes (1992) also used pulsed field electrophoresis to separate the chromosomes of the type culture of *Filobasidiella neoformans* var. *neoformans* and the tester strains (type a and type b). It was found that the two mating types had different karyotypes. It was demonstrated that type b was more virulent than type a in a murine virulence model with mating congenic strains. "These data suggested the genetic association of virulence with mating type in this fungal pathogen" (Kwon-Chung et al., 1992, p. 602). Wickes obtained his doctoral degree at Catholic University, Washington, DC, under the guidance of Kwon-Chung from the NIH and conducted his postdoctoral studies in Edman's laboratory at the University of California at San Francisco ("Training Trees", Appendix B).

The genetics of two virulence factors of *C. neoformans*, the capsule and melanin, were the focus of Eric Jacobson's research at the Veterans Hospital and MCV, VCU. Jacobson and colleagues (1982) examined the inheritance of the cryptococcal capsule, a virulence factor that had been described by Glenn Bulmer in the 1960s. The production of the phenotypic and genetic characteristics of capsule mutants were also reported by Jacobson's group. They concluded that "chromosomal genes are responsible for synthesis of the cryptococcal capsule and that these genes were not linked to auxotrophic markers" (p. 1292). By 1991, Jacobson and Herschell Emery had provided genetic support for a model that linked melanization with resistance to oxygen toxicity and virulence of *C. neoformans*. In 1993, Jacobson and Sheri Tinnell reported that the antioxidant capacity conferred by melanin had a value that approximated antimicrobial oxidant production by stimulated macrophages and they concluded that *C. neoformans* appeared to be protected by melanin against leukocyte oxidants. At University of Texas, Austin, Philip Geis, Michael Wheeler, and Paul Szaniszlo (1984) confirmed for the first time that melanin (a possible virulent factor) synthesis in *Wangiella dermatitidis* was produced by the pentaketide pathway. Mutants with a decrease or loss in melanin synthesis were induced and isolated.

Studies conducted at the University of Iowa by Bernice Slutsky, Jeffrey Buffo, and David Soll (1985) demonstrated that *C. albicans* switched heritably and reversibly at high frequency between several general phenotypes identified by colony morphology on agar. Based on these results, Soll and his group concluded that switching may provide *C. albicans* and other fungi with "the capacity to invade tissue and diverse body locations, evade the immune system, or change antibiotic resistance" (Slutsky et al., 1985, p. 669). In 1987, Soll, Carol Langtimm, Jan McDowell, James Hicks, and Rudolph Galask reported for the first time that "switching is occurring at the site of infection and provides a minimum estimate of the number of types of switching systems and strains which are involved in vaginal candidiasis" (p. 1611). In 1992, Brian Morrow, Thyagarajan Srikantha, and Soll identified the gene regulated by switching in *C. albicans*, which represented the transcript of an acid protease (*PEP1*) of the pepsinogen family. Two other genes were identified (Srikantha and Soll, 1993): *Op4*, which is on a different chromosome than *PEP1*, and the white-specific gene, *cWh11*, which is not transcribed in opaque cells.

Physiology and nutrition

Bruno Maresca et al. (1981) discovered that "cysteine stimulation of oxygen consumption is due to a cytosolic cysteine oxidase activity, which is present only in yeast-phase cells" (p. 4596). They proposed a model that outlined the unique biochemical events during the three stages of temperature-induced mycelial to yeast form transition of *H. capsulatum* var. *capsulatum*. The cysteine oxidase from the yeast cells of *H. capsulatum* was isolated and purified by Vijaya Kumar et al. (1983). This enzyme appeared to play an important role in the form conversion of *H. capsulatum* var. *capsulatum*. In 1993, Paul Szaniszlo and his students reported that an extremely low pH synthetic medium containing 0.1 mmol 1^{-1} Ca^{2+} induced large numbers of muriform cells of three etiologic agents of chromoblastomycosis (*Cladophialophora* [as *Cladosporium*] *carrionii*, *Fonsecaea pedrosoi*, and *Phialophora verrucosa*) as well as in the model of *Wangiella dermatitidis*. Leonel Mendoza, S. M. Karuppayil, and Szaniszlo (1993) indicated that "Ca_2 concentrations in human tissue may play a paramount role in the dimorphic switching between hyphae and sclerotic bodies" (p. 157). In the same year (1993), Chester Cooper and Szaniszlo developed an artificial parasexual cycle in *W. dermatitidis* and identified two of its cell division cycle genes, *CDC1* and *CDC2*. The role of these genes in pathogenesis was unclear, but they govern yeast bud emergence. Mendoza, Cooper, Peng and Karuppayil were doctoral and postdoctoral students of Szaniszlo; Szaniszlo's major professor was John Nathaniel Couch ("Training Trees", Appendix B). More recently with Peng and Cooper, the Szaniszlo group (personal communication, 1995) has molecularly transformed *W. dermatitidis* and subsequent students have cloned and disrupted a number of its genes, including its four chitin synthase genes.

The role of oxidative injury induced by amphotericin B was investigated in 1986 by Marcia Sokol-Anderson, Janina Brajtburg, and Gerald Medoff. It was common knowledge that amphotericin B and other polygene macrolide compounds bind to the cell membrane's ergosterol, a major sterol in fungal cells (cholesterol in mammalian cells). However, the mechanism of cytotoxicity was not well defined. Sokol-Anderson et al. (1986) described two unique mechanisms of action of amphotericin B on fungal cells, a pre-lethal leakage of potassium (low amphotericin B doses) and a lytic effect caused by the doses used clinically. The latter effect was the only one closely tied to oxidative damage. Sokol-Anderson was a Dalldorf Fellow (1985–1987) who was sponsored by Medoff at Washington University ("Training Trees", Appendix B).

Training and education contributions 1980 to 1996

Support from the NIH for medical mycology training has been very important to the discipline since the first award in 1958. By 1990, there were two sources of funding at the NIH, intramural and extramural. Intramural funding for medical mycology research was awarded to John Bennett and Kyung Joo Kwon-Chung at the National Institute of Allergy and Infectious Diseases (NIAID), to Philip Pizzo and Walsh at the NIH National Cancer Institute, and to the Infectious Diseases Section (T. Walsh, personal communication, March 15, 1994). The portfolio of extramural NIAID grants for medical mycology was administered by Dennis Dixon, beginning in 1992 when his position was created. Only 0.6% ($75 312, an average of $18 828 each) of the NIAID extramural funding for medical mycology was allocated for four training grants in the 1991 fiscal year. In contrast, immunology, genetics, and molecular biology research and clinical trials continued to receive most of the NIH funds. The NIAID extramural funding for medical mycology research amounted to $10 292 695 for 57 projects (an average of $180 574 each), or 89.1% of these NIAID funds (Bullock, Kozel, Scherer, and Dixon, 1993; CVs; questionnaires).

In 1996, a NIAID training grant (T32) was awarded to Montana State University with Jim Cutler as principal investigator (J. Cutler, personal communication, April 8, 1996). The intent of this interdisciplinary program, entitled Medical Mycology Predoctoral Training Program (MMPT), was to develop Ph.D. scientists who would pursue a research career in medical mycology. In addition to the NIH-funded training program at Montana State University, NIH has granted funds for medical mycology basic research through program project grants to the UCLA Harbor Medical Center (Principal Investigator, John Edwards), Georgetown University (Principal Investigator, Richard Calderone), and the University of California, San Diego (Principal Investigator, Theo Kirkland). These investigators focused their research activities on the development of a vaccine for coccidioidomycosis (Kirkland), the study of the pathogenesis of *Candida* (Calderone) as well as understanding the mechanisms of adherence of *Candida* cells to humans cells (Edwards). These centers are important resources for basic research-oriented postdoctoral training.

In the development and training of graduates in medical mycology, Judith Domer sees a shift from simple diagnosis to the study of basic mechanisms of infection and immunity. (J. Domer, personal communication, March 15, 1994). Errol Reiss sees a change in medical mycology focus since the 1980s as new areas emerge for training and research (E. Reiss, personal communication, January 25, 1994). Great emphasis is being placed on molecular biology, cloning and expression of genes, and the development of immunological testing, mostly with support from the pharmaceutical industry as funds from governmental sources have been reduced substantially. The transition in funding and support of medical mycology programs has changed the direction of the discipline since the early 1980s. Many of the active training centers and programs that were in operation in the 1960s and 1970s either ceased to exist or have been substantially reduced in scope. Important examples of centers that have closed are as follows.

The medical mycology program at Columbia University ceased to exist in 1981 at the time when Silva-Hutner retired (M. Silva-Hutner, personal communication, November 20, 1993; Irene Weitzman, personal communication, August 8, 1994). At MCV, VCU, student interest had waned by the mid-1980s, because the focus of the Microbiology Department was basic research. Jean Shadomy's medical mycology graduate course was discontinued in 1986 and it was replaced by a section of a microbiology course for graduate students given by the physician, Eric Jacobson, who trained in medical

mycology at Washington University. Jacobson's lectures concentrated on *C. neoformans* genetics and pathogenicity, the immunology of mycoses, rapid diagnosis, fungal metabolites and dimorphism (E. Jacobson, personal communication, January 4, 1994). Clinical mycology was taught to the medical students by Thomas Kerkering and Jacobson as part of grand rounds and intern training. Smith Shadomy's work in the field of antifungal susceptibility testing ended with his untimely death on June 22, 1992. Jean Shadomy retired from MCV, VCU the following year and Ana Espinel-Ingroff has been the director of the Medical Mycology Research Laboratory since 1990.

The medical mycology reference laboratory at the University of Kentucky was closed in 1982. When Michael Furcolow retired in 1984, Norman Goodman moved to the Department of Pathology and due to the difficulty in doing medical mycology research, student interest waned. This training program no longer exists (N. Goodman, personal communication, May 24, 1994). In 1988, Michael McGinnis moved from the University of North Carolina (UNC), Chapel Hill to the University of Texas Medical Branch at Galveston to create a medical mycology research center (McGinnis, 1994) where Chester Cooper and Lester Pasarell joined him. McGinnis' course was discontinued at UNC. Roy Hopfer continued informal medical mycology training at UNC via five hours of lectures to first-year medical students and maintained an active research laboratory. When Bill Cooper died in 1987, he was not replaced at Baylor University. The Naval Biological Laboratory in Oakland, California was closed in the early 1990s due to budget problems. (H. Levine, personal communication, June 27, 1994). Both Everett Beneke and Alvin Rogers retired in 1993 from Michigan State University and their training program was closed. However, Leonel Mendoza is now at this institution.

After Leanor Haley's retirement in 1986, the training branches at the CDC were discontinued because the federal government decided that the CDC should focus on the prevention and epidemiology of diseases. Before her death (May, 1996), Haley expressed great concern about reductions in medical mycology training programs in the United States (L. Haley, personal communication, December 31, 1993). Libero Ajello retired from the CDC after 43 years of service in September of 1990 after Lucille Georg and William Kaplan retired in the 1980s;

neither was replaced. The Molecular Biology Section continued to be an active reference point for molecular biology and genetics research (see scientific contributions sections) and informal training. In addition, training was provided by Arvind A. Padhye, Leo Kaufman, and Paul Standard on a one-to-one level at the CDC reference laboratory (A. Padhye, personal communication, March 16, 1994). Recently, the CDC group and the National Laboratory Training Network Southeastern Office began conducting workshops.

Conventional training approaches have been challenged and the need for more medical research has become paramount. Although training in medical mycology has not disappeared, it has taken a more informal direction under the guidance of individual researchers with a shift in emphasis towards one-on-one training. Another disturbing trend is that graduate schools are creating molecular biologists who work with fungi, but who do not know the biology of the fungal cell (I. Salkin, personal communication, February 22, 1994).

Based on the research conducted for this study (questionnaires, interview responses, CVs, and reviews of scientific papers), two avenues to training are discussed below: formal and informal. Formal training is given through structured courses or sections of courses at the university level for M.S. and Ph.D. students (often tied to teaching hospitals). Informal training is mostly associated with laboratory visits and, more importantly, postdoctoral training with individual researchers (often tied to teaching hospitals). Training from both avenues remains available at the institutions noted in the following sections. Since 1996, some of the individuals listed in the following sections have retired, moved to another institution, or died.

Formal training

Those centers having established medical mycology training programs that continued through the 1980s were recognized previously (See Chapters III to V). In addition, other centers also have provided formal training as described above (see "Training Trees", Appendix B). These centers and their principal investigator(s) are listed as follows: the University of Oklahoma at Oklahoma City (Juneann Murphy); the University of Iowa College of Medicine (David Soll); Wayne State University (Charles Jeffries); the University of Nevada (Thomas Kozel); Georgia State University at Atlanta (Donald Ahearn and Sally Meyers); the University of Texas at Austin (Paul Szaniszlo); the University of Iowa (David Soll and John Cazin); and the University of Minnesota College of Biological Sciences (Paul and Beatrice Magee). In most instances, informal training also is available at these institutions.

Judith Domer, a Tulane University graduate under Morris Shaffer, became Director of the Mycology Program at that institution in 1980 when Friedman retired. Although NIH training grants have been discontinued, Domer continued an active research program and was awarded NIH basic research grants for graduate and postdoctoral fellows (J. Domer's CV; "Training Trees", Appendix B). A medical mycology graduate course continues to be taught at Montana State University by Jim Cutler. Gene M. Scalarone, who trained with George Cozad at the University of Oklahoma and Hillel Levine at the University of California, Berkeley, came to the Idaho State University in 1980. Beginning in 1981, he taught a medical mycology course at this institution.

Informal training

Opportunities for informal training in medical mycology are quite diverse and include several new medical centers in addition to the ones described in previous chapters and Appendix B. Up to 1996, most of these centers specialized in postdoctoral training, including: the University of Iowa College of Medicine (Michael Pfaller); the University of Alabama (William Dismukes, Michael Sagg, and Peter Pappas); the VA Medical Center at Ann Arbor, Michigan (Carol Kauffman); the University of California at San Francisco (Jeffrey Edman); Myco Pharmaceuticals, Inc. at Cambridge, Massachusetts (William Timberlake); the UCLA Medical Center-Harbor (John Edwards and Scott Filler); and the University of Minnesota (Paul T. and Beatrice Magee).

Michael Rinaldi arrived in 1984 at the University of Texas, San Antonio, and initiated an active medical mycology reference laboratory for fungal identification and antifungal susceptibility testing. Rinaldi is well known for his numerous lectures and seminars, which he calls his "missionary work", that have promoted the discipline around the world. In addition, Rinaldi has provided workshops in clinical microbiology at Montana State University for trainees enrolled in the NIH-sponsored program under the direction of Cutler at that university. Kevin Hazen, a student of Cutler and George Kobayashi and Gerald Medoff ("Training Trees", Appendix B), is at the University of Virginia Medical Center (UVA) in Charlottesville as Associate Professor of Pathology and Microbiology. Trainees from Cutler's program have the opportunity to spend two to three months of clinical mycology laboratory training under Hazen's direction at UVA.

At the University of Cincinnati College of Medicine, Cincinnati, Ohio, many individuals have dedicated their careers to medical mycology since Jan Schwarz and Gerald Baum began their pioneering work with *H. capaulatum* var. *capsulatum* in the late 1940s. Other outstanding medical mycologists who established medical mycology laboratories and provided postdoctoral training at this university are the physicians Ward Bullock, who trained George Deepe (Dalldorf Fellow, 1981–1983; "Training Trees", Appendix B), his successor; Brian Wong; and P. Walzer. The Ph.D. scientist Judith Rhodes came to this university following her postdoctoral training (Dalldorf Fellow, 1980–1982, "Training

Trees", Appendix B). Wong, who had come to this university in 1984, moved to the VA Connecticut Health Care Center in West Haven and the Yale University School of Medicine. He received his medical mycology postdoctoral training (1978–1980) under the guidance of the physician Donald Armstrong at the Memorial Sloan-Kettering Cancer Center and Cornell University Medical College.

Elizabeth Keath conducted her postdoctoral medical mycology training between 1985 and 1990 with Kobayashi and Medoff and established a basic research laboratory at Saint Louis University, St. Louis, Missouri. Her laboratory provides another source of medical mycology training ("Training Trees", Appendix B).

An innovative collaborative effort entitled, "Focus on Fungal Infections", was initiated in 1990 by Elias Anaissie and Michael Rinaldi. The purpose of this annual conference "is the continued provision of practical and updated information to those clinicians and laboratorians who are involved in the care of patients with invasive mycoses" (Conference program). The conference assembles internationally recognized experts to discuss the latest information on the diagnosis and management of a wide range of fungal pathogens in humans.

VII.   Summary and conclusions

This historical study has traced the development of medical mycology as a discipline in the United States for the past 100 years. The discipline began within the context of sweeping technological and social changes in this country beginning in the late 1890s. Not long before, the country had emerged from the trauma and destruction of the Civil War, followed by westward expansion, as pioneers shook off the emotional depression of that terrible conflict. After the Civil War, the standard of living increased rapidly and brought demands for a better quality of life, including medical care and treatment. Known technologies were improved, such as the microscope, and bacteriological laboratories were established in the two newly founded medical schools at Columbia University in New York and the Johns Hopkins University and Hospital in Baltimore. The stage was set for the study of fungal diseases and the immediate reaction of the medical community to diagnose and treat these newly realized threats to public health.

The epoch designated as "The Era of Discovery" encompasses the period from the late 1800s to 1919. The newly established medical center at Johns Hopkins, coupled with the creation of a "modern" bacteriological research laboratory, enabled a thorough investigation of the diseases caused by fungi to be initiated. Publications on fungal diseases coming from Europe and Latin America gave American investigators comparative studies and a scientific reference point, which were other contributing factors.

In the late 1890s and early 1900s, research rapidly expanded, findings were published and presented at scientific meetings, and fungal diseases began to be considered in the differential diagnosis of disease processes. As a consequence, other fungal diseases soon were described which led to the search for diagnostic tools in the laboratory and the initiation of studies to define and manage mycotic diseases. Although the discipline was established, research was isolated, there was much confusion, and no organized training or education was available; only a few physicians were concerned about the prevalence of mycoses which they considered to be rare, but fatal.

During "The Formative Years" (1920–1949), the development of the discipline was advanced by the establishment of medical mycology foci of learning and dissemination of knowledge at universities and government supported reference centers (1926 to 1947). Fortuitously, the decision to establish medical mycologic research and training programs occurred when several trained, competent, young investigators were just graduating. Although early medical mycologists had little knowledge of the fungi pathogenic to humans, they used the biology of plant fungal pathogens as a foundation to build upon.

Rhoda Benham, highly regarded within the discipline as a teacher and scientist, trained the first generation of medical mycologists in the United States and Chester Emmons established the federal government's first medical mycology research and reference laboratory at the NIH. This center developed into a fundamental source for the creation and dissemination of basic and clinical medical mycology information; it also had a profound impact on directing the field. Confusion regarding fungal nomenclature was approached by developing logical classification systems and distinguishing descriptions of many fungal diseases. These contributions were turning points for the discipline, because they greatly improved the scientific knowledge of these pathogens.

World War II brought a massive infusion of funds and the best technology to cope with the high incidence of fungal infections in military personnel. Medical mycology became significant as nearly every important leader in the field either served as a consultant and trainer during that time or received much of their orientation and training from those early leaders. Immediately following the war, the demand for training led Norman Conant, another early key figure, to develop his famous medical mycology textbook, summer course, and his graduate study program. The impact of these training and educational activities resulted in more and better trained personnel to deal with the diagnosis of fungi in the clinical laboratory and improved patient management. The foundation for the next generation of medical mycologists had been established.

Among the individuals influenced and or trained by Conant were Howard Larsh and Libero Ajello, who became important leaders in the field. The higher incidence of mycoses during the war and the recognition that fungal diseases were not rare led the CDC to establish the division for medical mycology. Ajello was chosen to initiate a CDC reference laboratory in Atlanta in 1947 and he formed a team of medical mycologists who developed important laboratory diagnostic tests, conducted practical short courses for physicians and laboratory personnel, and disseminated needed applied research information to the nation. The CDC approach became a model for applied research, reference laboratory work, and training.

The period between 1950 and 1969 is designated in this study as "The Advent of Antifungal and Immunosuppressive Therapies". Although the first antifungal agents were discovered in the late 1940s, and bacterial infections were treated with specific drugs since the 1930s, effective chemotherapy was not available for the treatment of systemic fungal infections in humans at the end of the 1940s. When the 1950s began, the mortality rate for patients with invasive mycoses stood at 50–100%, dependent on the severity of the disease, which demanded improved treatment. The direction of the discipline was changed in 1955, when patients with invasive fungal diseases were treated successfully with the newly developed amphotericin B.

For the first time, effective antifungal agents were developed for patient management. This was followed by a new wave of opportunistic fungal infections as a result of the use of immunosuppressive chemotherapy for oncology and transplantation patients. The government and the pharmaceutical communities provided the financial, technological, and human resources that brought rapid advancements in the field. Competent, well trained clinicians and researchers were insufficient in number to meet the higher incidence of severe and fatal opportunistic mycoses being diagnosed in immunosuppressed patients. The NIH, the CDC, and other government sponsored centers initiated evaluations of antifungal therapies and developed more, improved, and faster diagnostic tests, respectively. In 1968, the demand for more medical mycology training led to the appointment of Leanor Haley at the CDC, as the Chief of the newly founded Medical Mycology Training Branch with the mandate to expand and further develop training opportunities across the country. Others within the discipline saw the need to expand academic education and initiated the transition from studies of the biology of fungi to studies at the molecular level.

Professional mycological groups were influential in developing medical mycology, because they provided a unified voice on issues, gave peer recognition, initiated peer review to upgrade standards in the field, and promoted the discipline. The establishment of a Medical Mycology Division within ASM and MMSA brought together individuals as a group that could express concerns and speak on behalf of the field with respect to needs, priorities, and a vision for the future.

The decade of the 1970s comprises what this study defines as "The Years of Expansion". This was the period when the science flourished and expanded at all levels of training and research. Two major scientific events changed and enlarged the role of fungal diseases in clinical medicine. First, there was an increase in the spectrum of available antibiotics and their use; second, more immunosuppressive and cytotoxic therapeutic agents were developed and utilized. The latter rendered patients highly susceptible to opportunistic fungal infections. Due to the increased incidence of these infections, rapid tests and commercial fungal identification systems were developed and refined, while existing methodologies were improved. The availability of these new tests stimulated interest in mycology among microbiologists and physicians and by simplifying diagnosis, patient care was improved.

The wider use of antibacterial, cytotoxic, and immunosuppressive chemotherapy created additional demands for advanced formal and informal medical mycology training, as well as research. Particularly acute was the demand for trained clinical laboratory personnel and the need to provide physicians with concise and up-to-date information on the fungal pathogens. As a result, informal training programs for large and varied audiences began at a number of universities and medical centers. The NIH, private companies and foundations responded with additional funds to support training and research programs during "the Years of Expansion".

Unfortunately, the need for partnerships was not always evident because there were sufficient resources available for investigators and trainers to work independently. The Brown-Hazen funds that came from the royalties set aside from the discovery of nystatin by Rachel Brown and Elizabeth Hazen were devoted primarily to the development of structured medical mycology training programs that included research. Grants from the NIH and Brown-Hazen Foundation were essential to the continuation and development of new postdoctoral and graduate training programs during the 1970s.

The understanding of cell defense and other immunologic mechanisms, virulence, and genetic processes was viewed as a means to develop new strategies for the control and prevention of fungal diseases. Host–parasite interaction studies became more important as infectious diseases, including mycoses, were commonly diagnosed among immunocompromised patients. In the pursuit of a complete understanding of fungal pathogens, investigations on the genetics, immunochemistry, and taxonomy of the fungi were intensified. More importantly, these advances created the training and research resources that began to replace the structured academic training programs that were terminated during the next era.

"The Era of Transition" describes the years from 1980 to 1994. This period also could be called the era of contrasts because several critical, yet conflicting, elements came to the forefront. The first cases of AIDS in the early 1980s introduced a new population of immunosuppressed patients, who were high risks for severe, opportunistic fungal infections. Although it would appear that the field of medical mycology should have received more funding support in response to this threat, funding resources were either cut substantially by the federal government or ended when private foundation support was depleted. This resulted in the termination of sponsored NIH training programs at Washington University, Tulane University, the University of California at Los Angeles, the University of Kentucky, and the University of Oklahoma.

Coupled with the reduction in financial resources, this era may be characterized by several trends in the transition of leadership. Teams formed at important centers, including the CDC, were disbanded or terminated because a number of early medical mycology leaders dedicated to training and applied research either retired or died. At the same time, new and larger collaborative groups from different medical centers were formed by both clinical and basic scientists. Many of these individuals either focused in certain areas of medical mycology or were specialists from other disciplines who came into the field

and for whom medical mycology was only a part of their professional work. The result was a shift in leadership from traditionally trained medical mycologists to a greater number of scientists that became interested in the study of mycoses and their etiologic agents.

Because of these trends, laboratory diagnostic tests and epidemiological studies were improved significantly by the application of DNA-based methodology and numerous other technological advances. Both intramural and extramural NIH grants were awarded for clinical and basic studies, whereas training grants were decreased or eliminated. Intramural support for medical mycology continued at the NIH under the guidance of the clinical investigators and basic scientists. Also, NIH grants and corporate support were needed for clinical evaluations of drug therapy and other applied research investigations that were initiated at additional medical centers.

Difficulty in the management of new opportunistic infections, especially in AIDS patients, spurred greater dependence on pharmaceutical and corporate funds. Companies that were developing the alternative oral imidazole and triazole antifungal agents in the United States and Europe provided financial support for clinical trials. In the late 1970s, the NIH began to sponsor the NIAID MSG group, an important collaborative group of physicians. These two sources of revenue made possible the joint evaluations of the new oral antifungal agents developed during the 1970s as well as reevaluation of existing ones. Collaborative studies established therapeutic regimens for the management of certain severe fungal infections in both AIDS and non-AIDS patients.

In 1996, important conclusions from the research for this book were as follows:

1. The future direction of medical mycology is endangered by what may be considered a crisis in training as a result of substantial reductions in federal funds and a depletion of private resources. There was a transition from established, formal training programs to research and informal training resources at isolated research laboratories under the guidance of principal investigators. Medical mycology courses as complete units have nearly ceased to exist, and education in the discipline currently is

being limited to sections in microbiology courses and lectures for medical students. The crisis in medical mycology training is more evident when it is recognized that there is a dearth of educators and researchers with the knowledge and experience to address the practical problems of the science. In addition, there is a decline in the number and quality of diagnostic laboratories and trained personnel to staff them.

2. The remaining years of the 1990s appear to present significant issues for medical mycologists. Parallel to the crisis in training, medical mycology is advancing as a discipline as a consequence of important research contributions from a more diversified group of scientists coming into the field. These contributions made possible the shifting to molecular biology and genetics research, which appear to be necessary to combat the increased virulence and the resistance of pathogenic fungi to chemotherapeutic agents. While these new technologies are being applied to the discipline, the cost of equipment and the dearth of properly trained personnel may limit their widespread use. The demand for targeting resources towards basic research and applied investigations that improve patient care, limits support for organized training programs in medical mycology. These trends suggest that the current status of the discipline may be compared with its early years of uncertainty.

3. Increased public pressure for cost containment and accountability for expenditures narrows the focus of research activities. As a result of the restrictions to funding resources, the creativity of basic and applied researchers is being channeled toward goal oriented studies. Because of this, investigators have become more specialized and must, of necessity, form partnerships and collaborative groups.

4. The methodologies for education also have been significantly altered. Rapid developments in electronic media are replacing laboratory training under the guidance of skilled investiga-

tors. How to keep clinical laboratory personnel in step with these rapid changes has become a major challenge. It is costly and time consuming to equip and properly train the investigators and staff needed, but the demand for their services at many institutions does not support such investment. The result is fewer laboratories that have the resources necessary for complete diagnostic work.

5. Further, the rapidity of change places uncertainties on the discipline as to what constitutes proper education in medical mycology, even if more resources were available. What should leaders in the field and officials at higher education institutions and medical schools do in response to the current state of the discipline? Suggested areas of study include the evaluation of baseline competency in medical mycology as well as appropriate educational models to prepare individuals for their various roles. Investigators should consider the continuing effect of technology and the availability of reliable sources of funds. Also, it is important to take into account emerging trends regarding the incidence and prevalence of fungal infections for further educational and scientific advancement that will lead to improvements in patient care.

As medical mycologists prepare to address these uncertainties, there is a critical need to combine proven and effective training with the new technologies. Organizations in control of resources must recognize the importance of basic research advances without neglecting the practical aspects of the field, especially training and education. Ongoing funding is necessary for both applied and basic medical mycology education and research to thrive.

J.H. Robinson (1912) stated that the more we learn about the past of any movement, program, or institution, the more we will be able to understand the nature of the problems and the prospects for that institution. That is, the more we know how anything became the way it is, the better prepared we will be to propose solutions and to chart future courses of action.

References

Ainsworth GC. Raimond Sabouraud (1864–1938). Sabouraudia 1961; 1: 1.

Ainsworth GC. Introduction to the History of Medical and Veterinary Mycology. Cambridge: Cambridge University Press, 1986.

Ajello L, Zeidberg LD. Isolation of *Histoplasma capsulatum* and *Allescheria boydii* from soil. Science 1951; 113: 662–663.

Ajello L, Georg LK. *In vitro* hair cultures for differentiating between atypical isolates of *T. mentagrophytes* and *T. rubrum*. Mycopathologia 1957; 8: 3–17.

Ajello L, Cheng S-L. The perfect state of *Trichophyton mentagrophytes*. Sabouraudia 1967; 5: 230–234.

Ajello L. Chronological Record of the Medical Mycological Society of the Americas' Genesis. In: Morris Gordon, ed. Bull Med Mycological Soc Am 1968; 5: 1–2.

Ajello L. Establishment of the Rhoda Benham Award. In: Morris Gordon, ed. Bull Med Mycological Soc Am 1969; 8: 1–2.

Ajello L, Georg LK, Steigbigel RT, Wang CJK. A case of phaeohyphomycosis caused by a new species of *Phialophora*. Mycologia 1974; 66: 490–498.

Ajello L. Milestones in the history of medical mycology: the dermatophytes. In: K Iwataz, ed. International Society for Human and Animal Mycology, Recent Advances in Medical and Veterinary Mycology. Tokyo: University Tokyo, 1975, 3–11.

Ajello L. Lucille K. Georg-Pickard – Obituary. In: Morris Gordon, ed. Bull Med Mycological Soc Am 1981; 37: 3–4.

American Society for Microbiology. Primary divisional voting recap, Archives Collection, The Library. Baltimore County: University of Maryland, 1973.

Annals of New York Academy of Science. First Medical Mycology Monograph 1950; 50: 1209–1404.

Aronson JD, Saylor RM, Parr EI. Relationship of coccidioidomycosis to calcified pulmonary nodules. Arch Pathol 1942; 34: 31–48.

Artz RP, Bullock WE. Immunoregulatory responses in experimental disseminated histoplasmosis: Depression of T-cell-dependent and T-effector responses by activation of splenic suppressor cells. Infect Immun 1979; 23: 893–902.

Ashworth JH. On *Rhinosporidium seeberi* (Wernicke 1903) with special reference to its sporulation and affinities. Trans Roy Soc Edinburgh 1923; 53: 301–342.

Backus MP, Evans RI. HC Greene (1904–1967). Mycologia 1968; 60: 994–998.

Bacon WS. Elizabeth Lee Hazen, 1885–1975. Mycologia 1976; 68: 961–969.

Baker RD. Tissue changes in fungous disease. Arch Pathol 1947; 44: 459–466.

Baldwin RS. The Fungus Fighters and Their Discovery. Ithaca and London: Cornell University Press, 1981.

Bauer H, Ajello L, Adams E, Hernandez DU. Cerebral mucormycosis: Pathogenesis of the disease. Am J Med 1955; 18: 822–831.

Bauer H, Flanagan JF, Sheldon WH. Experimental cerebral mucormycosis in rabbits with alloxan diabetes. Yale J Biol Med 1955; 28: 29–36.

Beaman L, Pappagianis D, Benjamini E. Significance of T cells in resistance to experimental murine coccidioidomycosis. Infect Immun 1977; 17: 580–585.

Beaman L, Benjamini E, Pappagianis D. Activation of macrophages by lymphokines: Enhancement of phagosome-lysosome fusion and killing of *Coccidioides immitis*. Infect Immun 1983; 39: 1201–1207.

Beeson PB. One hundred years of American internal medicine. Ann Int Med 1986; 105: 436–444.

Beneke ES, Wilson RW, Rogers AL. Extracellular enzymes of *Blastomyces dermatitidis*. Mycopathologia 1969; 39: 325–328.

Benham RW. Certain monilias parasitic on man: Their identification by morphology and by agglutination. J Infect Dis 1931; 49: 183–215.

Benham RW. The fungi of blastomycosis and coccidioidal granuloma. Arch Dermatol Syphil (Chicago) 1934; 30: 385–400.

Benham RW. Cryptococci – Their identification by morphology and by serology. J Infect Dis 1935; 57: 255–274.

Benham RW. The cultural characteristics of *Pityrosporum ovale* – a lipophilic fungus. J Invest Dermatol 1939; 2: 187–202.

Benham, RW, Georg LK. *Allescheria boydii*, causative agent in a case of meningitis. J Invest Dermatol 1948; 10: 99–110.

Benham, RW. Cryptococcosis and blastomycosis. Ann NY Acad Sci 1950a; 50, 1299–1314.

Benham RW. The Laboratory of Medical Mycology. The Biopsy. 2: 1. New York: Department of Dermatology, Columbia-Presbyterian Medical Center, 1950b.

Benham RW. Nutritional studies of the dermatophytes – effect on growth and morphology, with special reference to the production of macroconidia. Trans NY Acad Sci 1953; 15, 102–106.

Bennett JE, Dismukes WE, Duma RJ et al. A comparison of amphotericin B alone and combined with flucytosine in the treatment of cryptococcal meningitis. N Engl J Med 1979; 301: 126–131.

Berg FT. Om torsk hos Barn. Stockholm: LJ Hjerta, 1846.

Berkhout CM. De Schimmelgeschlachten Monilia, Oidium, Oospora, en Torula. Dissertation, University of Utrecht, 1923.

Berliner MD, Reca ME. Release of protoplasts in the yeast phase of *Histoplasma capsulatum* without added enzyme. Science 1970;167: 1255–1257.

Bernard EM, Wong B, Armstrong D. Stereoisomeric configuration of arabinitol in serum, urine, and tissues in invasive candidiasis. J Infect Dis 1985; 151: 711–715.

Bille J, Stockman L, Roberts GD, Horstmeier CD, Ilstrup DM. Evaluation of a lysis-centrifugation system for recovery of yeasts and filamentous fungi from blood. J Clin Microbiol 1983; 18: 469–471.

Binazzi M. Italian memoirs of Aldo Castellani. Int J Dermatol 1991; 30: 741–745.

Binford CH, Thompson RK, Gorham ME, Emmons CW. Mycotic brain abscess due to *Cladosporium trichoides*, a new species. Am J Clin Pathol 1952; 22: 535–542.

Blanchard DK, Michelini-Norris MB, Djeu JY. Production of granulocyte-macrophage colony-stimulating factor by large granular lymphocytes stimulated with *Candida albicans*: Role in activation of human neutrophil function. Blood 1991; 79: 2259–2265.

Block ER, Jennings AE, Bennett JE. 5-Fluorocytosine resistance in *Cryptococcus neoformans*. Antimicrob Agents Chemother 1973; 3: 647–656.

Bloomfield N, Gordon MA, Elmendorf Jr, DF. Detection of *Cryptococcus neoformans* antigen in body fluids by latex particle agglutination. Proc Soc Exp Biol Med 1963; 114: 64–67.

Bodey GP. Fungal infections complicating acute leukemia. J Chronic Dis 1966; 19: 667–687.

Bowman WB. Coccidioidal granuloma. Am J Roentgen 1919; 6: 547–555.

Boyd MF, Crutchfield ED. A contribution to the study of mycetoma in North America. Am J Trop Med 1921; 1: 215–289.

Bozzette SA, Larsen RA, Chiu J, Leal MA, Jacobsen J, and the California Collaborative Treatment Group. A placebo-controlled trial of maintenance therapy with fluconazole after treatment of cryptococcal meningitis in the acquired immunodeficiency syndrome. N Engl J Med 1991; 324: 580–584.

Brummer E, Stevens DA. Activation of murine polymorphonuclear neutrophils for fungicidal activity with supernatants from antigen-stimulated immune spleen cell cultures. Infect Immun 1984; 45: 447–542.

Brummer E, Kurita N, Yoshida S, Nishimura K, Miyaji M. A basis for resistance of *Blastomyces dermatitis* killing by human neutrophils: Inefficient generation of myeloperoxidase system products. J Med Vet Mycol 1992; 30: 233–243.

Brummer E, Bhagavathula PR, Hanson LH, Stevens DA. Synergy of itraconazole with macrophages in killing *Blastomyces dermatitidis*. Antimicrob Agents Chemother 1992; 35: 2487–2492.

Brummer E, Hanson LH, Stevens DA. IL-4, IgE, and interferon-γ production in pulmonary blastomycosis: Comparison in mice untreated, immunized, or treated with an antifungal (SCH 39304). Cell Immunol 1993; 149: 258–267.

Brumpt E. Precis de Parasit, 3rd edn. Paris: Masson et Cie, p. 1105, 1922.

Buchman TG, Rossier M, Merz WG, Charache P. Detection of surgical pathogens by in vitro DNA amplification. Part I. Rapid identification of *Candida albicans* by in vitro amplification of a fungus-specific gene. Surgery 1990; 108: 338–347.

Bullock W, Kozel T, Scherer S, Dixon DM. Medical mycology in the 1990s: Involvement of NIH and the wider community. Am Soc Microbiol News, 1993; 59(4), 182–185.

Bulmer GS, Sans MD, Gunn CM. *Cryptococcus neoformans*. I. Nonencapsulated mutants. J Bacteriolology 1967; 94: 1475–1479.

Bulmer GS, Sans MD. *Cryptococcus neoformans*. II. Phagocytosis by human leukocytes. J Bacteriol 1967; 94: 1480–1483.

Burke RC. Tinea versicolor: Susceptibility factors and experimental infection in human beings. J Invest Dermatol 1961; 36: 389–402.

Burnie JP, Williams JD. Evaluation of the Ramco latex agglutination test in the early diagnosis of systemic candidiasis. Eur J Clin Microbiol 1985; 4: 98–101.

Buschke A. Ueber eine durch coccidien Hervergerufene Krankheit des menschen. Deutsche med Wochenschr 1895; 21(3): 14.

Busey JF, Baker R, Birch L et al. Blastomycosis. I. A review of 198 collected cases in Veterans Administration Hospitals. Am Rev Resp Dis 1964; 89: 659–672.

Busse O. Ueber parasitare zelleinschlüsse und ihre züchtung. Zentralbl Bakterial 1894; 16: 175–180.

Calderone RA, Linehan L, Wadsworth E, Sandberg AL. Identification of C3d receptors on *Candida albicans*. Infect Immun 1988; 56: 252–258.

Campbell CC. Reverting *Histoplasma capsulatum* to the yeast phase. J Bacteriol 1947; 54: 263–264.

Campbell C. Chester Wilson Emmons – "DEAN of U.S. Medical Mycologists". Unpublished manuscript from Medical Mycological Society of the Americas Archives, 1983.

Campbell C. Norman Francis Conant – "The Beloved Instructor". Unpublished manuscript from Medical Mycological Society of the Americas Archives, 1984.

Caplan LM, Merz WG. Evaluation of two commercially prepared biphasic media for recovery of fungi from blood. J Clin Microbiol 1978; 8: 469–470.

Carle GF, Olson MV. Separation of chromosomal DNA molecules from yeasts by orthogonal-field alternation gel electrophoresis. Nucleic Acids Res 1984; 12: 5647–5664.

Carle GF, Frank M, Olson MV. Electrophoretic separations of large DNA molecules by periodic inversion of the electric field. Science 1986; 232: 65–68.

Caroline L, Rosner F, Kozinn PJ. Elevated serum iron, low unbound transferrin and candidiasis in acute leukemia. Blood 1969; 34: 441–451.

Carter KC. The development of Pasteur's concept of disease causation and the emergence of specific causes in nineteenth-century medicine. Bull Hist Med 1991; 65: 528–548.

Centers for Disease Control. Kaposi's sarcoma and *Pneumocystis* pneumonia among homosexual men, New York City and California. Morbid Mortal Weekly Rep 1981; 30: 305–308.

Cerqueira-Pinto AGC. Keratomycose nigricans palmar. Tése Fac Med Bahia, Brazil, 1916.

Chernin E. A unique tribute to Theobald Smith, 1915. Rev Infect Dis 1987; 9: 625–635.

Chezzi C. Dr. Libero Ajello (editorial). Eur J Epidemiol 1992; 8(3): 319–320.

Chick EW, Evans J, Baker RD. The inhibitory effect of amphotericin B on localized *Rhizopus oryzae* infection (mucormycosis) utilizing the pneumoderma pouch of the rat. Antibiot Chemother 1958; 8: 506–510.

Chick EW, Furcolow ML. Second Annual Report of the Mycology Program. Lexington: University of Kentucky and VA Hospital, 1971.

Christie A, Peterson JC. Pulmonary calcification in negative reactors to tuberculin. Am J Public Health, 1945; 35: 1131–1147.

Chu G, Vollrath D, Davis RW. Separation of large DNA molecules by contour-clamped homogeneous electric fields. Science 1986; 234: 1582–1585.

Cole GT, Saha K, Seshan KR et al. Retrovirus-induced immunodeficiency in mice exacerbates gastrointestinal candidiasis. Infect Immun 1992; 60: 4168–4178.

Cole GT, Zhu S, Hsu L, Kruse D, Seshan KR, Wang F. Isolation and expression of a gene which encodes a wall-associated proteinase of *Coccidioides immitis*. Infect Immun 1992; 60: 416–427.

Conant NF. A bit of history, third annual MMSA lecture. In: Morris Gordon, ed. Bull Med Mycological Soc Am 1969; 7: 1–4.

Converse JL. Effect of surface active agents on endosporulation of *Coccidioides immitis* in a chemically defined medium. J Bacteriol 1957; 74: 106–107.

Cooke JV. Immunity tests in coccidioidal granuloma. Arch Int Med 1915; 15: 479–486.

Coons AA, Kaplan MH. Localization of antigen in tissue cells. II. Improvements in a method for the detection of antigen by means of fluorescent antibody. J Exp Med 1950; 91: 1–13.

Cooper CR, Szaniszlo PJ. Evidence for two cell division cycle (CDC) genes that govern yeast bud emergence in the pathogenic fungus *Wangiella dermatitidis*. Infect Immun 1993; 61: 2069–2081.

Cox RA, Britt LA. Antigenic identity of biologically active antigens in coccidioidin and spherulin. Infect Immun 1987; 55: 2590–2596.

Creitz J, Harris HW. Isolation of *Allescheria boydii* from sputum. Am Rev Tuberc 1955; 71: 126–130.

Currie BP, Freundlich LF, Casadevall A. Restriction fragment length polymorphism analysis of *Cryptococcus neoformans* isolates from environmental (pigeon excreta) and clinical sources in New York City. J Clin Microbiol 1994; 32: 1188–1192.

Cutler JE, Glee PM, Horn HL. *Candida albicans*- and *Candida stellatoidea*-specific DNA fragment. J Clin Microbiol 1988; 26: 1720-1724.

Da Fonseca O, de Arêa Leão AC. Sobre os cogumelos da piedra Brasileira. Mem Inst Oswaldo Cruz suppl das Memorias, no. 4: desde 1928.

Da Rocha-Lima H. Beitrag zur kenntnis der Blastomykoses Lymphangitis epizootica und Histoplasmosia. J Zentralbl Bakteriol 1912–1913; 67: 233–249.

Darling ST. A protozoön general infection producing pseudotubercles in the lungs and focal necroses in the liver, spleen and lymphnodes. J Am Med Assoc, 1906; 46: 1283–1285.

Davidson AM, Gregory PH. In situ cultures of dermatophytes. Can J Res 1934; 10: 373–393.

Davis DJ. The morphology of *Sporothrix schenckii* in tissues and in artificial media. J Infect Dis 1913; 12: 453–458.

de Hoog GS, Guého E, Masclaux F, Gerrits van den Ende AHG, Kwon-Chung KJ, McGinnis MR. Nutritional physiology and taxonomy of human-pathogenic *Cladosporium-xylohypha* species. J Med Vet Mycol 1995; 33: 339–347.

Deepe GS. Protective immunity in murine histoplasmosis: Functional comparison of adoptively transferred T-cell clones and splenic T-cells. Infect Immun 1988; 56: 2350–2355.

De Monbreun WA. The cultivation and culturation characteristics of Darling's *Histoplasma capsulatum*. Am J Trop Med 1934; 14: 93–125.

Denning DW, Tucker RM, Hanson LH, Stevens DA. Treatment of invasive aspergillosis with itraconazole. Am J Med 1989; 86: 791–800.

Denning DW, Tucker RM, Hanson LH, Hamilton JR, Stevens DA. Itraconazole therapy for cryptococcal meningitis and cryptococcosis. Arch Intern Med 1989; 149: 2301–2308.

Denning DW, Lee JY, Hostetler JS et al. NIAID mycoses study group multicenter trial of oral itraconazole therapy for invasive aspergillosis. Am J Med 1994; 97: 135–144.

Denton JF, McDonough ES, Ajello L, Ausherman RJ. Isolation of *Blastomyces dermatitidis* from soil. Science 1961; 133: 1126–1127.

Derensinski S, Hector R. The history of Coccidioidomycosis. In: Coccidioidomycosis. Proceedings of the 5th International Conference. Bethesda, MD: National Foundation for Infectious Diseases, 1996.

de Repentigny L, Marr LD, Keller JW et al. Comparison of enzyme immunoassay and gas-liquid chromotography for the rapid diagnosis of invasive candidiasis in cancer patients. J Clin Microbiol 1985; 21: 972–979.

Diamond RD, Bennett JE. Prognostic factors in cryptococcal meningitis. A study in 111 cases. Ann Intern Med 1974; 80: 176–180.

Diamond RD, Allison AC. Nature of the effector cells responsible for antibody dependent cell-mediated killing of *Cryptococcus neoformans*. Infect Immun 1976; 14: 716–720.

Diamond RD. Effects of stimulation and suppression of cell-mediated immunity on experimental cryptococcosis. Infect Immun 1977; 17: 187–194.

Diamond RD, Krzesicki R. Mechanisms of attachment of neutrophils to *Candida albicans* pseudohyphae in the absence of serum, and of subsequent damage to pseudohyphae by microbicidal processes of neutrophils in vitro. J Clin Invest 1978; 61: 360–369.

Diamond RD, Krzesicki R, Epstein B, Jao W. Damage to hyphal forms of fungi by human leukocytes in vitro: A possible host defense mechanism in aspergillosis and mucormycosis. Am J Pathol 1978; 91: 313–323.

Dickson EC. Oidiomycosis in California, with especial reference to coccidioidal granuloma. Arch Intern Med 1915; 16: 1028–1044.

Dickson EC. "Valley Fever" of the San Joaquin Valley and fungus *coccidioides*. California Western Med 1937a; 47: 151–155.

Dickson EC. *Coccidioides* infection I. Arch Intern Med 1937b; 57: 1029–1044.

Dickson EC, Gifford MA. *Coccidioides* infection (coccidioidomycosis). II. The primary type of infection. Arch Intern Med 1938; 62: 853–871.

Dismukes WE, Cloud G, Gallis HA, Kerkering TM, Medoff G and the National Institute of Allergy and Infectious Diseases Mycoses Study Group. Treatment of cryptococcal meningitis with combination amphotericin B and flucytosine for four as compared with six weeks. N Engl J Med 1987; 317: 334–341.

Dismukes WE, Bradsher RW, Cloud GC et al. Itraconazole therapy for blastomycosis and histoplasmosis. Am J Med 1992; 93: 489–497.

Djeu JY, Blanchard DK, Halkias D, Friedman H. Growth inhibition of *Candida albicans* by human polymorphonuclear neutrophils: Activation by interferon-γ and tumor necrosis factor. J Immunol 1986; 137: 2980–2984.

Dodd K, Tompkins EH. A case of histoplasmosis of Darling in an infant. Am J Trop Med 1934; 14: 127–137.

Dolman CE. Theobald Smith (1859–1934), pioneer American microbiologist. Perspect Biol Med 1982; 25(3): 417–427.

Domer JE. Monosaccharide and chitin content of cell walls of *Histoplasma capsulatum* and *Blastomyces dermatitidis*. J Bacteriol 1971; 207: 870–877.

Domer J, Friedman L. Billy H. Cooper (1936–1987). In: Morris Gordon, ed. Bull Med Mycological Soc Am 1988; 54: 8–9.

Domer JE, Garner RE, Befidi-Mengue RN. Mannan as an antigen in cell-mediated immunity (CMI) assays and as a modulator of mannan-specific CMI. Infect Immun 1989; 57: 693–700.

Dong ZM, Murphy JW. Mobility of human neutrophils in response to *Cryptococcus neoformans* cells, culture filtrate antigen, and individual components of the antigen. Infect Immun 1993; 61: 5067–5077.

Dorn GL, Land GA, Wilson GE. Improved blood culture technique based on centrifugation: Clinical evaluation. J Clin Microbiol 1979; 9: 391–396.

Drouhet E. Education and training in medical mycology with an introduction to medical mycology. In: DK Arora, L Ajello KG Mukerji. eds. Handbook of Applied Mycology: Vol. 2, Humans, Animals, and Insects. New York: Marcel Dekker, 1992: 1–49.

Drutz DJ, Spickard A, Rogers DE, Koenig MG. Treatment of disseminated mycotic infections. A new approach to amphotericin B therapy. Am J Med 1968; 45: 405–418.

Drutz D, Frey CL. Intracellular and extracellular defenses of human phagocytes against *Blastomyces dermatitidis* conidia and yeasts. J Lab Clin Med 1985; 105: 737.

Edman JC, Kwon-Chung KJ. Isolation of the URA5 gene from *Cryptococcus neoformans* var. *neoformans* and its use as a selective marker for transformation. Mol Cell Biol 1990; 10: 4538–4544.

Elson WO. The antibacterial and fungistatic properties of propamidine. J Infect Dis 1945; 76: 193–197.

Emerson R, Humber RA. Robert Meredith Page February 5, 1919–May 17, 1968. Mycologia 1970; 62: 1085–1093.

Emmons CW. Dermatophytes. Natural grouping based on the form of the spores and accessory organs. Arch Dermatol Syphilol (Chicago) 1934; 30: 337–362.

Emmons CW. Isolation of *Coccidioides* from soil and rodents. Public Health Rep Washington, 1942; 57: 109–111.

Emmons CW. *Allescheria boydii* and *Monosporium apiospermum*. Mycologia 1944; 36: 188–193.

Emmons CW. *Phialophora jeanselmei* comb. n. from mycetoma of the hand. Arch Pathol 1945; 39: 364–368.

Emmons CW. Isolation of *Histoplasma capsulatum* from soil. Public Health Reports, Washington, 1949; 64: 892–896.

Emmons CW. Isolation of *Cryptococcus neoformans* from soil. J Bacteriol 1951; 62: 685-690.

Emmons CW. Saprophytic sources of *Cryptococcus neoformans* associated with the pigeon (*Columba livia*). Am J Hygiene 1955; 62: 227–232.

Emmons CW. The elusive fungi, second annual MMSA lecture. In: Morris Gordon, ed. Bull Med Mycological Soc Am 1968; 4: 1–3.

Ernst HC. A case of mucor infection. J Med Res 1918; 39: 143–146.

Espinel-Ingroff A, Kish Jr CW, Kerkering TM et al. Collaborative comparison of broth macrodilution and microdilution antifungal susceptibility tests. J Clin Microbiol 1992; 30: 3138–3145.

Espinel-Ingroff A. Medical mycology in the United States: 100 years of development as a discipline, Ph.D. dissertation. Virginia Commonwealth University, Richmond, 1994.

Espinel-Ingroff A. History of medical mycology in the United States. Clin Microbiol Rev, 1996; 9: 235–272.

Evans AS. Causation and disease: Effect of technology on postulates of causation. Yale J Biol Med 1991; 64: 513–528.

Evans EE. An immunologic comparison of twelve strains of *Cryptococcus neoformans* (*Torula histolytica*). Proc Soc Exp Biol Med 1949; 71, 644-646.

Evans EE. The antigenic composition of *Cryptococcus neoformans*. I. A serologic classification by means of the capsular and agglutination reactions. J Immunol 1950; 64: 423–430.

Evans EE, Kessel JF. The antigenic composition of *Cryptococcus neoformans*. II. Serologic studies with the capsular polysaccharide. J Immunol 1951; 67: 109–114.

Evans EE, Theriault RJ. The antigenic composition of *Cryptocccus neoformans*. IV. The use of paper chromatography for following purification of the capsular polysaccharide. J Bacteriol 1953; 65: 571–577.

Eveland WC, Marshall JD, Silverstein AM, Johnson FB, Iverson L, Wilson DJ. Specific immunochemical staining of *Cryptococcus neoformans* and its polysaccharide in tissue. Am J Pathol 1957; 33: 616–617.

Fidel PL, Murphy JW. Effects of cyclosporin A on the cells responsible for the anticryptococcal cell-mediated immune response and its regulation. Infect Immun 1989; 57: 1158–1164.

Fiese MJ. Treatment of disseminated coccidioidomycosis with amphotericin B: report of a case. California Med 1957; 86: 119–124.

Fineman BC. A study of the thrush parasite. J Infect Dis 1921; 28: 185–200.

Foerster HR. Sporotrichosis, an occupational dermatosis. J Am Med Assoc 1926; 87: 1605–1609.

Francis P, Lee JW, Hoffman A et al. Efficiency of unilamellar liposomal amphotericin B in treatment of pulmonary aspergillosis in persistently granulocytopenic rabbits: The potential role of bronchoalveolar D-mannitol and serum galactomannan as markers of infection. J Infect Dis 1994; 169(2): 356–368.

Freeman W, Weidman FD. Cystic blastomycosis of the cerebral gray matter. Arch Neurol Psychiatr 1923; 2: 589–603.

Freeman W. Fungus infections of the central nervous system. Ann Intern Med 1933; 6: 595–607.

Fromtling RA, Galgiani JN, Pfaller MA et al. Multicenter evaluation of a broth macrodilution antifungal susceptibility test for yeasts. Antimicrob Agents Chemother 1993; 37: 39–45.

Frothingham L. Lesion in the lung of a horse. A tumor-like lesion in the lung of a horse caused by a blastomyces (*Torula*). J Med Res 1902; 8: 31–43.

Fuchs A. Fifty years Antonie van Leeuwenhoek, its history and its impact. Antonie van Leeuwenhoek 1984; 50: 425–432.

Furcolow ML. Undated autobiography, submitted by N. Goodman for Medical Mycological Society of the Americas Archives.

Furcolow ML. Comparison of treated and untreated severe histoplasmosis. A Communicable Disease Center Cooperative Mycoses Study. J Am Med Assoc 1963; 183: 823–829.

Furcolow ML, Chick EW, Busey JF, Menges RW. Prevalence and incidence studies of human and canine blastomycosis. Am Rev Respir Dis 1970; 102: 60–67.

Galgiani JN, Isenberg RA, Stevens DA. Chemotaxigenic activity of extracts from the mycelial and spherule phases of *Coccidioides immitis* for human polymorphonuclear leukocytes. Infect Immun 1978; 21: 862–865.

Galgiani JN, Catanzaro A, Cloud GA et al. Fluconazole therapy for coccidioidal meningitis. Ann Intern Med 1993; 119: 28–35.

Gantz NM, Swain JL, Medeiros AA, O'Brien TF. Vacuum blood-culture bottles inhibiting growth of *Candida* and fostering growth of *Bacteroides*. Lancet 1974; 2: 1174–1176.

Gass RS, Gauld RL, Harrison EF, Stewart HC, Williams WC. Tuberculosis studies in Tennessee. Am Rev Tuberc 1938; 38: 441–447.

134

Geiger AJ, Wenner HA, Axilrod HD, Durlacher SH. Mycotic endocarditis and meningitis. Yale J Biol Med 1946; 18: 259–268.

Geis PA, Wheeler MH, Szaniszlo PJ. Pentaketide metabolites of melanin synthesis in the dematiaceous fungus *Wangiella dermatitidis*. Arch Microbiol 1984; 137: 324–328.

Gentles JC. Experimental ringworm in guinea pigs: Oral treatment with griseofulvin. Nature 1958; 182: 476–477.

Gentry LO, Wilkinson ID, Lea AS, Price MF. Latex agglutination test for detection of *Candida* antigen in patients with disseminated disease. Eur J Clin Microbiol 1983; 2: 122–128.

Georg LK. Influence of nutrition on growth and morphology of the dermatophytes. Trans NY Acad Sci 1949; 11: 281–286.

Georg LK, Ajello L, Gordon MA. A selective medium for the isolation of *Coccidioides immitis*. Science 1951; 114: 387–389.

Georg LK, Camp LB. Routine nutritional tests for the identification of dermatophytes. J Bacteriol 1957; 74: 113–121.

Georg LK, Ajello L, Friedman L, Brinkman SA. A new species of *Microsporum* pathogenic to man and animals. Sabouraudia 1962; 1: 189–196.

Gilchrist TC. Protozoan dermatitidis. J Cutan Dis 1894; 12: 496.

Gilchrist TC. A case of blastomycetic dermatitis in man. Johns Hopkins Hosp Rep 1896; 1: 269–283.

Gilchrist TC, Stokes WR. The presence of an *Oidium* in the tissues of a case of pseudo-lupus vulgaris. Preliminary report. Bull Johns Hopkins Hosp 1896; 7: 129–133.

Gilchrist TC, Stokes WR. A case of pseudo-lupus vulgaris caused by a *Blastomyces*. J Exp Med 1898; 3: 53–83.

Gold W, Stout HA, Pagano JF, Donovick R. Amphotericin A and B, antifungal antibiotics produced by a streptomycete. I. In vitro studies of A. Antibiot Annu, 1955–1956, NY Med Encyclopedia, pp. 579–586.

Gomori G. A new histochemical test for glycogen and mucin. Am J Clin Pathol 1946; 16: 177–179.

Goodman NL. Biography of Michael L Furcolow. Unpublished manuscript from Medical Mycological Society of the Americas Archives, 1985.

Goodman NL, Roberts GD. Howard W Larsh – 1914–1993. Mycopathologia 1994; 125: 1–2.

Gordon MA. The lipophilic mycoflora of the skin. I. In vitro culture of *Pityrosporum orbiculare*. n.sp. Mycologia 1951; 43: 524–535.

Gordon MA. Differentiation of yeasts by means of fluorescent antibody. Proc Soc Exp Biol Med 1958; 92: 694–698.

Gordon MA, Lapa E. Serum protein enhancement of antibiotic therapy in cryptococcosis. J Infect Dis 1964; 114: 373–377.

Gordon MA. A Career in Medical Mycology (his recollection of training and research). Unpublished manuscript from Medical Mycological Society of the Americas Archives, 1993.

Graybill JR, Stevens DA, Galgiani JN et al. Ketoconazole treatment of coccidioidal meningitis. Ann NY Acad Sci 1988; 544: 488–496.

Graybill JR, Stevens DA, Galgiani JN, Dismukes WE, Cloud GA, and the NIAID-MSG group. Itraconazole treatment of coccidioidomycosis. Am J Med 1990; 89: 282–290.

Gregory JE, Golden A, Haymaker W. Mucormycosis of the central nervous system. A report of three cases. Bull Johns Hopkins Hosp 1943; 73: 405–419.

Gridley MF. A stain for fungi in tissue sections. Am J Clin Pathol 1953; 23: 303–307.

Grocott RC. A stain for fungi in tissue and smears using Gomori's methenamine-silver nitrate technique. Am J Clin Pathol 1955; 25: 975–979.

Hageage GJ, Harrington BJ. Use of calcofluor white in clinical mycology. Lab Med 1984; 15: 109–112.

Haggerty TE, Zimmerman LE. Mycotic keratitis. S Med J 1958; 51: 153–159.

Halde C. Course Manual, Mycology: New Viewpoints. San Francisco, CA: University of California at San Francisco, 1971.

Halde C. Course Manual, Opportunistic Mycoses of AIDS, Leukemia, and Other Compromised Patients: Mycology Laboratory Aspects. San Francisco, CA: University of California at San Francisco, 1987.

Hall GS, Pratt-Rippin K, Washington JA. Evaluation of a chemiluminescent probe assay for identification of *Histoplasma capsulatum* isolates. J Clin Microbiol 1992; 30: 3003–3004.

Hamburger WW. A comparative study of four strains of organisms isolated from four cases of generalized blastomycosis. J Infect Dis 1907; 4: 201–209.

Hansemann D von. Über eine bisher nicht beobachtete Gehirnerkrankung durch Hefen. Verh Dtsch Ges Path 1905; 9: 21–24.

Hansmann GH, Schenken JR. A unique infection in man caused by a new yeast-like organism, a pathogenic member of the genus *Sepedonium*. Am J Pathol 1934; 10: 731–738.

Harrell ER, Curtis AC. The treatment of North American blastomycosis with amphotericin B. Arch Dermatol 1957; 76: 561–569.

Hasenclever HF, Mitchell WO. Antigenic studies of *Candida*. J Bacteriol 1961; 82: 578–581.

Hawksworth DL. The fungal dimension of biodiversity: Magnitude, significance, and conservation. Mycological Res 1991; 95(6): 641–655.

Hazen EL. *Microsporum audouinii*: The effect of yeast extract, thiamine, pyridoxine, and *Bacillus weidmaniensis* on the colony characteristics and macroconidial formation. Mycologia 1947; 39: 200–209.

Hazen EL, Brown R. Two antifungal agents produced by a soil actinomycete. Science 1950; 112: 423.

Heiner DC. Diagnosis of histoplasmosis using precipitin reactions in agar gel. Pediatrics 1958; 22: 616–627.

Hektoen L, Perkins CF. Refractory subcutaneous abscesses caused by *Sporothrix schenckii*. A new pathogenic fungus. J Exp Med 1900–1901; 5: 77–89.

Henrici AT. An endotoxin from *Aspergillus fumigatus*. J Immunol 1939; 36: 319–338.

Hesseltine CW. Dorothy I. Fennell. Mycologia 1979; 71: 889–891.

Hidore MR, Nabavi N, Sonleitner F, Murphy JW. Murine natural killer cells are fungicidal to *Cryptococcus neoformans*. Infect Immun 1991; 59: 1747–1754.

Hirsch EF, Benson H. Specific skin and testis reactions with culture filtrates of *Coccidioides immitis*. J Infect Dis 1927; 40: 629–633.

Hodges RS. Cultures of ringworm fungi on Sabouraud's proof mediums and on mediums prepared with American peptones and sugars. Arch Dermatol Syphilol 1928; 18: 852–856.

Hopfer RL, Gröschel D. Six-hour pigmentation test for the identification of *Cryptococcus neoformans*. J Clin Microbiol 1975; 2: 96–98.

Hopfer RL, Mills K, Gröschel D. Improved blood culture medium for radiometric detection of yeasts. J Clin Microbiol 1979; 9: 448–449.

Hopfer RL, Walden P, Setterquist S, Highsmith WE. Detection and differentiation of fungi in clinical specimens using polymerase chain reaction (PCR) amplification and restriction enzyme analysis. J Med Vet Mycol 1993; 31: 65–75.

Horta P. Sobre una nova forma de piedra. Mem Inst Oswaldo Cruz 1911; 3: 87–88.

Horta P. Sobre um caso de tinha preta e um novo cogumelo (*Cladosporium werneckii*). Rev Med Cirug Brazil 1921; 29: 269–274.

Hotchkiss RD. A microchemical reaction resulting in the staining of the polysaccharide structures in fixed tissue preparations. Arch Biochem 1948; 16: 131–141.

Howard DH. Dimorphism in *Sporotrichum schenckii*. J Bacteriol 1961; 81: 464–469.

Howard DH. Intracellular growth of *Histoplasma capsulatum*. J Bacteriol 1965; 89: 518–523.

Howard DH. In appreciation. In: M. Gordon, ed. Bull Med Mycological Soc Am 1974; 23: 4–5.

Howard DH, Otto V. Experiments on lymphocyte-mediated cellular immunity in murine histoplasmosis. Infect Immun 1977; 16: 226–231.

Howard DH. Biographical Sketches on JR Kessel, OA Plunkett, JW Wilson (Contribution No. 88 of the Collaborative California Universities Mycology Unit). Los Angeles: University of California at Los Angeles, 1985.

Hube B, Turner CJ, Odds FC et al. Sequence of the *Candida albicans* gene encoding the secretory aspartate proteinase. J Med Vet Mycol 1991; 22: 129–132.

Huppert M, Keeney EL. Immunization against superficial fungous infection. J Dermatol 1959; 52: 15–19.

Huppert M, Bailey JW. The use of immunodiffusion test in coccidioidomycosis. The accuracy and reproducibility of the immunodiffusion which correlates with complement fixation. J Clin Pathol 1965; 44: 364–368.

Huppert M, Peterson ET, Sun SH, Chitjian PA, Derrevere WJ. Evaluation of a latex particle agglutination test for coccidioidomycosis. Am J Clin Pathol 1968; 49: 96–102.

Huppert M, Oliver DJ, Sun SH. Combined methenamine silver nitrate and hematoxylin eosin stain for fungi in tissues. J Clin Microbiol 1978; 8: 598–603.

Hutter RVP, Lieberman PH, Collins HS. Aspergillosis in a cancer hospital. Cancer 1964; 17: 747–756.

Jacobson ES, Ayers DJ, Harrell AC, Nicholas CC. Genetic and phenotypic characterization of capsule mutants of *Cryptococcus neoformans*. J Bacteriol 1982; 150: 1292–1296.

Jacobson ES, Emery HS. Catecholamine uptake, melanization, and oxygen toxicity in *Cryptococcus neoformans*. J Bacteriol 1991; 173: 401–403.

Jacobson ES, Tinnell SB. Antioxidant function of fungal melanin. J Bacteriol 1993; 175: 7102–7104.

Joachim H, Polayes SH. Subacute endocarditis and systemic mycosis (*Monilia*). J Am Med Assoc 1940; 115: 205–208.

Johnson SM, Zimmermann CR, Pappagianis D. Amino-terminal sequence analysis of the *Coccidioides immitis* chitinase/immunodiffusion-complement fixation protein. Infect Immun 1993; 61: 3090–3092.

Kanbe T, Li R-K, Wadsworth E, Calderone RA, Cutler JE. Evidence for expression of the C3d receptor of *Candida albicans* in vitro and in vivo obtained by immunofluorescence and immunoelectron microscopy. Infect Immun 1991; 59: 1832–1838.

Kaplan W, Ivens MS. Fluorescent antibody staining of *Sporotrichum schenckii* in cultures and clinical materials. J Invest Dermatol 1960; 35: 151–159.

Kaplan W, Kraft DE. Demonstration of pathogenic fungi in formalin-fixed tissues by immunofluorescence. Am J Clin Pathol 1969; 52: 420–432.

Kass EH. History of the speciality of infectious diseases in the United States. Ann Intern Med 1987; 106: 745–756.

Kass EH, Hayes KM. A history of the Infectious Diseases Society of America. Rev Infect Dis 1988; 10(2): 1–159.

Kauffman CA, Israel KS, Smith JW, White AC, Schwarz J, Brooks GF. Histoplasmosis in immunosuppressed patients. Am J Med 1978; 64: 923–932.

Kaufmann CS, Merz WG. Electrophoretic karyotypes of *Torulopsis glabrata*. J Clin Microbiol 1989; 27: 2165–2168.

Kaufman L, Kaplan W. Preparation of a fluorescent antibody specific for the yeast phase of *Histoplasma capsulatum*. J Bacteriol 1961; 82: 729–735.

Keath EJ, Spitzer ED, Painter AA, Travis SJ, Kobayashi GS, Medoff G. DNA probe for the identification of *Histoplasma capsulatum*. J Clin Microbiol 1989; 27: 2369–2372.

Keddie F, Shadomy S. Etiological significance of *Pityrosporum orbiculare* in tinea versicolor. Sabouraudia 1963; 3: 21–25.

Keller HW. Travis E. Brooks (1917–1976). Mycologia 1979; 71: 233–237.

Kennedy Jr, TJ. The rising cost of NIH-funded biomedical research? Acad Med 1990; 65: 63–73.

Kerkering TM, Duma RJ, Shadomy S. The evolution of pulmonary cryptococcosis. Ann Intern Med 1981; 94: 611–616.

Kessler G, Nickerson WJ. Glucomannan-protein complexes from cell walls of yeasts. J Biol Chem 1959; 234: 2281–2285.

Keye JD Jr, Magee WE. Fungal diseases in a general hospital. Am J Clin Pathol 1956; 26: 1235–1253.

Khakpour FR, Murphy JW. Characterization of a third-order suppressor T cell (Ts3) induced by cryptococcal antigen(s). Infect Immun 1987; 55: 1657–1662.

Kiehn TE, Bernard EM, Gold JWM, Armstrong D. Candidiasis: Detection by gas-liquid chromatography of D-arabinitol, a fungal metabolite, in human serum. Science 1979; 206: 577–580.

Kiessling R, Klein E, Wigzell H. "Natural" killer cells in the mouse. I. Cytotoxic cells with specificity for mouse Moloney leukemia cells. Specificity and distribution according to genotype. Eur J Immunol 1975; 5: 112.

King DF, King LAC. George Thin (1839–1903): Pioneer British dermatopathologist and mycologist. Am J Dermatopathol 1988; 10: 546–548.

Kirkland TN, Zhu S, Kruse D, Hsu L, Seshan KR, Cole GT. *Coccidioides immitis* fractions which are antigenic for immune T lymphocytes. Infect Immun 1991; 59: 3952–3961.

Kirsh B. Forgotten Leaders in modern medicine. Valentin, Gruby, Remak, Auerbach (Includes: X. Hube, 1837, Doctoral dissertation, DeMorboscrofuloso). Trans Am Philos Soc 1954; 44: 139–317.

Klein RS, Harris CA, Small CB, Moll B, Lesser M, Friedland GH. Oral candidiasis in high-risk patients as the initial manifestation of the acquired immunodeficiency syndrome. N Engl J Med 1984; 311: 354–358.

Kligman AM, Mescon H, DeLamater ED. The Hotchkiss-McManus stain for the histopathologic diagnosis of fungus diseases. Am J Clin Pathol 1951; 21: 86–91.

Kligman AM. The pathogenesis of tinea capitis due to *Microsporum audouinii* and *Microsporum canis*. I. Gross observa-

136

tions following the inoculation of humans. J Invest Dermatol 1952; 18: 231–246.

Kligman AM. Tinea capitis due to *M. audouinii* and *M. canis*. II. Dynamics of the host-parasite relationship. Arch Dermatol Syphilol 1955; 71: 313–337.

Kobayashi G. Summary of Training Proposal. St. Louis: Washington University, 1982.

Kobayashi G. Personal Reflections on Significant Medical Mycology Books. Unpublished manuscript from Medical Mycological Society of the Americas Archives, 1993.

Komorowski RA, Farmer SG. Rapid detection of candidemia. Am J Clin Pathol 1973; 59: 56–61.

Kozel TR, McGaw TG. Opsonization of *Cryptococcus neoformans* by human immunoglobulin G: Role of immunoglobulin G in phagocytosis by macrophages. Infect Immun 1979; 25: 255–261.

Kozel TR, Wilson MA, Pfrommer GST, Schlageter AM. Activation and binding of opsonic fragments of C3 on encapsulated *Cryptococcus neoformans* by using an alternative complement pathway reinstituted from six isolated proteins. Infect Immun 1989; 57: 1922–1927.

Kruse D, Cole GT. A seroreactive 120-kilodalton β-1, 3-glucanase of *Coccidioides immitis* which may participate in spherule morphogenesis. Infect Immun 1992; 60: 4350–4363.

Kumar V, Maresca B, Sacco M, Goewert R, Kobayashi G, Medoff G. Purification and characterization of a cysteine dioxygenase from the yeast phase of *Histoplasma capsulatum*. Biochemistry 1983; 22: 762–768.

Kunin CM. E Pluribus Unum. J Infect Dis 1988; 157: 400–404.

Kurtz MB, Cortelyou MW, Kirsch DR. Integrative transformation of *Candida albicans*, using a cloned *Candida* ADE2 gene. Mol Cell Biol 1986; 6: 142–149.

Kwon KJ, Fennell DI, Raper KB. A heterothallic species of *Aspergillus*. Am J Bacteriol 1964; 51: 679.

Kwon-Chung KJ. *Emmonsiella capsulata*: Perfect stage of *Histoplasma capsulatum*. Science 1972; 177: 368–369.

Kwon-Chung KJ. Perfect state (*Emmonsiella capsulata*) of the fungus causing large-form African histoplasmosis. Mycologia 1975a; 47: 980–990.

Kwon-Chung KJ. A new genus, *Filobasidiella*, the perfect state of *Cryptococcus neoformans*. Mycologia 1975b; 57: 1197–1200.

Kwon-Chung KJ, Lehman D, Good C, Magee PT. Genetic evidence for role of extracellular proteinase in virulence of *Candida albicans*. Infect Immun 1985; 49: 571–575.

Kwon-Chung KJ, Campbell CC. Chester Wilson Emmons. J Med Vet Mycol 1986; 24: 89–90.

Kwon-Chung KJ, Edman JC, Wickes BL. Genetic association of mating types and virulence in *Cryptococcus neoformans*. Infect Immun 1992; 60: 602–605.

Kyle RA, Shampo MA. Agostino Bassi. J Am Med Assoc 1979; 241: 1584.

Land GA, Vinton EC, Adcock GB, Hopkins JM. Improved auxanographic method for yeast assimilations: A comparison with other approaches. J Clin Microbiol 1975; 2: 206–227.

Land GA, McDonald WC, Stjernholm RL, Friedman L. Factors affecting filamentation in *Candida albicans*: Changes in respiratory activity of *Candida albicans* during filamentation. Infect Immun 1975; 12: 119–127.

Land GA, Harrison BA, Hulme KL, Cooper BH, Byrd JC. Evaluation of the new API 20C strep for yeast identification against a conventional method. J Clin Microbiol 1979; 3: 357–364.

Lane CG. A cutaneous lesion caused by a new fungus (*Phialophora verrucosa*). J Cutan Dis 1915; 33: 840–846.

Langenbeck B. Auffindung von pilzen auf der scheim haut der speiseröhre einer typhus-leiche. Froriep's neue notizen 1839; 252: 145–147.

Leach BE, Ford JH, Whiffen AJ. Actidione, an antibiotic from *Streptomyces griseus*. J Am Chem Soc 1947; 69: 474.

Lehmann PF, Reiss E. Invasive aspergillosis: Antiserum for circulating antigen produced after immunization with serum from infected rabbits. Infect Immun 1978; 20: 570–572.

Lehmann PF, Lin D, Lasker BA. Genotypic identification and characterization of species and strains within the genus *Candida* by using random amplified polymorphic DNA. J Clin Microbiol 1992; 30: 3249–3254.

Lehrer RI, Cline MJ. Interaction of *Candida albicans* with human leukocytes and serum. J Bacteriol 1969; 98: 996–1004.

Lehrer RI, Ganz T, Szklarek D, Selsted ME. Modulation of the in vitro candidacidal activity of human neutrophil defensins by target cell metabolism and divalent cations. J Clin Invest 1988; 81: 1829–1835.

Levine HB, González-Ochoa A, Ten Eyck DR. Dermal sensitivity to *Coccidioides immitis*. A comparison of responses elicited in man by spherulin and coccidioidin. Am Rev Respir Dis 1973; 107: 379–386.

Levitz SM, Diamond RD. Mechanisms of resistance of *Aspergillus fumigatus* conidia to killing by neutrophils in vitro. J Infect Dis 1985; 152: 33–42.

Levitz SM, Selsted ME, Ganz T, Lehrer RI, Diamond RD. In vitro killing of spores and hyphae of *Aspergillus fumigatus* and *Rhizopus oryzae* by rabbit neutrophil cationic peptides and bronchoalveolar macrophages. J Infect Dis 1986; 154: 483–489.

Lim TS, Murphy JW, Cauley LK. Host-etiological agent interactions in intranasally and introperitoneally induced cryptococcosis in mice. Infect Immun 1980; 29: 633–641.

Littman ML, Horowitz PL. Coccoidioidomycosis and its treatment with amphotericin B in man. Am J Med 1958; 24: 568–592.

Littman ML. Cryptococcosis (Torulosis). Am J Med 1959; 27: 976.

Lockwood WR, Allison F, Batson BE, Busey JF. The treatment of North American blastomycosis. Ten years of experience. Am Rev Respir Dis 1969; 100: 314–320.

Lott TJ, Boiron P, Reiss E. An electrophoretic karyotype for *Candida albicans* reveals large chromosomes in multiples. Mol Gen-Genetic 1987; 209: 170–174.

Lott TJ, Page LS, Boiron P, Benson J, Reiss E. The nucleotide sequence of the *Candida albicans* aspartyl proteinase gene. Nucleic Acids Res 1989; 17: 1779.

Lott TJ, Kuykendall RJ, Welbel SF, Pramanik A, Lasker BA. Genomic heterogeneity in the yeast *Candida parapsilosis*. Curr Genet 1993; 23: 463–467.

Louria DB, Feder N, Emmons CW. Amphotericin B in experimental histoplasmosis and cryptococcosis. Antibiot Annu 1956–57: 870–877.

Louria DB. Some aspects of the absorption, distribution, and excretion of amphotericin B in man. Antibiot Med 1958; 5: 295–301.

Louria DB, Fallon N, Browne HG. The influence of cortisone on experimental fungus infections in mice. J Clin Invest 1960; 39: 1435–1449.

Louria DB, Brayton RG. Behavior of *Candida* cells within leukocytes. Proc Soc Exp Biol Med 1964; 115: 93–98.

Lupan DM, Cazin J Jr. Pathogenicity of *Allescheria boydii* for mice. Infect Immun 1973; 8: 743–751.

Lupan DM, Cazin J Jr. Serological diagnosis of petriellidiosis (allescheriosis). Mycopathologia 1976; 58: 31–38.

Lupan DM, Nziramasanga P. Collagenolytic activity of *Coccidioides immitis*. Infect Immun 1986; 51: 360–361.

Lurie HI, Still WJS. The "capsule" of *Sporotrichum schenckii* and the evolution of the asteroid body. A light and electron microscopic study. Sabouraudia 1969; 7: 64–70.

Lütz A. Uma mycose pseudosoccidioidica localizada a boca e observada no Brasil. Contribuicao ao conhecimento das hyphoblastomycoses americanas. Brasil-Méd 1908; 22: 121–124.

Magee BB, Magee PT. Electrophoretic karyotypes and chromosome numbers in *Candida* species. J Gen Microbiol 1987; 133: 425–430.

Magee BB, Koltin Y, Gorman JA, Magee PT. Assignment of cloned genes to the seven electrophoretically separated *Candida albicans* chromosomes. Mol Cell Biol 1988; 8: 4721–4726.

Magee BB, Hube B, Wright RJ, Sullivan PJ, Magee PT. The genes encoding the secreted aspartyl proteinases of *Candida albicans* constitute a family with at least three members. Infect Immun 1993; 61: 3240–3243.

Manning M, Mitchell TG. Morphogenesis of *Candida albicans* and cytoplasmic proteins associated with differences in morphology, strain, or temperature. J Bacteriol 1980; 144: 258–273.

Maresca B, Medoff G, Schlessinger D, Kobayashi GS. Regulation of dimorphism in the pathogenic fungus *Histoplasma capsulatum*. Nature 1977; 266: 447–448.

Maresca B, Lambowitz AM, Kumar VB, Grant GA, Kobayashi GS, Medoff G. Role of cysteine in regulating morphogenesis and mitochondrial activity in the dimorphic fungus *Histoplasma capsulatum*. Proc Nat Acad Sci 1981; 78: 4596–4600.

Martin DS, Baker RD, Conant NF. A case of verrucous dermatitis caused by *Hormodendrum pedrosoi*. Am J Trop Med 1936; 16: 593–619.

Martin AR. The systematic treatment of dermatophytoses (Letter). The Vet Rec 1958; 70: 1232.

Mason AB, Brandt ME, Buckley HR. Enolase activity associated with a *C. albicans* cytoplasmic antigen. Seventh International Symposium on Yeasts, S231–S239, 1989.

Mason MM, Lasker BA, Riggsby WS. Molecular probe for identification of medically important *Candida* species and *Torulopsis glabrata*. J Clin Microbiol 1987; 25: 563–566.

Mayo Clinic. Clinical Microbiology Residency Program. Rochester, MN. Undated brochure.

McDonough ES, Lewis AL. *Blastomyces dermatitidis*: Production of the sexual stage. Science 1967; 156: 528–529.

McDonough ES, Lewis A.L. The ascigerous stage of *Blastomyces dermatitidis*. Mycologia 1968; 60: 76–83.

McGinnis MR. Human pathogenic species of *Exophiala*, *Phialophora*, and *Wangiella*. Proceedings of the Fourth International Conference on the Mycoses 1977; 356: 37–59.

McGinnis MR, Katz B. *Ajellomyces* and its synonym *Emmonsiella*. Mycotaxon 1979; 8: 157–164.

McGinnis R, Padhye AA, Ajello L. *Pseudallescheria* Negroni et Fischer, 1943, and its later synonym *Petriellidium* Malloch, 1970. Mycotaxon 1982; 14: 94–102.

McGinnis MR. Chromoblastomycosis and phaeohyphomycosis: New concepts, diagnosis, and mycology. J Am Acad Dermatol 1983; 8: 1–16.

McGinnis MR, Rinaldi MG, Winn RE. Emerging agents of phaeohyphomycosis: Pathogenic species of *Bipolaris* and *Exserohilum*. J Clin Microbiol 1986; 24: 250–259.

McGinnis MR, Borelli D, Padhye AA, Ajello L. Reclassification of *Cladosporium bantianum* in the genus *Xylohypha*. J Clin Microbiol 1986; 23: 1148–1151.

McGinnis M. Autobiographical highlights of his career. Unpublished manuscript from Medical Mycological Society of the Americas Archives, 1994.

McKinsey DS, Kauffman CA, Pappas PG. Fluconazole therapy for histoplasmosis. Clin Infect Dis 1996; 23: 996–1001.

McManus JFA. Histological and histochemical uses of periodic acid. Stain Technol 1948; 23: 94–108.

Medlar EM. A cutaneous infection caused by a new fungus, *Phialophora verrucosa*, with a study of the fungus. J Med Res 1915; 32: 507–521.

Medoff G, Comfort M, Kobayashi GS. Synergistic action of amphotericin B and 5-fluorocytosine against yeast-like organisms (35943). Proc Soc Exp Biol Med 1971; 138: 571.

Medoff G, Kobayashi GS, Kwan CN, Schlessinger D, Venkov P. Potentiation of rifampicin and 5-fluorocytosine as antifungal antibiotics by amphotericin B. Proc Nat Acad Sci 1972; 69: 196–199.

Mendoza L, Karuppayil SM, Szaniszlo PJ. Calcium regulates in vitro dimorphism in chromoblastomycotic fungi. Mycoses 1993; 36: 157–164.

Meyer RD, Young LS, Armstrong D, Yu B. Aspergillosis complicating neoplastic disease. Am J Med 1973; 54: 6–15.

Meyer W, Mitchell TG, Freedman EZ, Vilgalys R. Hybridization probes for conventional DNA fingerprinting used as single primers in the polymerase chain reaction to distinguish strains of *Cryptococcus neoformans*. J Clin Microbiol 1993; 31: 2274–2280.

Miller BL, Miller KY, Timberlake WE. Direct and indirect gene replacement in *Aspergillus nidulans*. Mol Cell Biol 1985; 5: 1714–1721.

Miller GG, Witwer MW, Braude AI, Davis CE. Rapid identification of *Candida albicans* septicemia in man by gas-liquid chromatography. J Clin Invest 1974; 54: 1235–1240.

Mitchell T. Need for training in medical mycology (Training Grant Proposal). Durham: Duke University, 1991.

Monson TP, Wilkinson KP. Mannose in body fluids as an indicator of invasive candidiasis. J Clin Microbiol 1981; 14: 557–562.

Montgomery FH, Ormsby OS. Systemic blastomycosis. Arch Intern Med 1908; 2: 1–41.

Montoya y Flores JB. Recherches sur les Caratés de Colombie. These Med. Paris 1898; 25: 48–49.

Moore M, Ackerman LV. Sporotrichosis with radiate formation in tissue. Arch Dermatol Syphilol 1946; 53: 253–264.

Morrison CJ, Hurst SF, Bragg SL, Kuykendall RJ, Pohl H, Reiss E. Heterogeneity of the purified extracellular asparty proteinase from *Candida albicans*: Characterization with monoclonal antibodies and N-terminal amino acid sequence analysis. Infect Immun 1993; 61: 2030–2036.

Morrow B, Srikantha T, Soll DR. Transcription of the gene for a pepsinogen, PEP 1, is regulated by white-opaque switching in *Candida albicans*. Mol Cell Biol 1992; 12: 2997–3005.

Mosley RL, Murphy JW, Cox RA. Immunoabsorption of *Cryptococcus* – specific suppression T-cell factors. Infect Immun 1986; 51: 844–850.

138

Murphy JW, Gregory JA, Larsh HW. Skin testing of guinea pigs and footpad testing of mice with a new antigen for detecting delayed hypersensitivity to *Cryptococcus neoformans*. Infect Immun 1974; 9: 404–409.

Murphy JW, McDaniel DO. In vitro reactivity of natural killer (NK) cells against *Cryptococcus neoformans*. J Immunol 1982; 128: 1577–1583.

Murphy JW, Mosley RL, Cherniak R, Reyes GH, Kozel TR, Reiss E. Serological, electrophoretic, and biological properties of *Cryptococcus neoformans* antigens. Infect Immun 1988; 56: 424–431.

Murphy JW. Cytokine profiles associated with induction of the anticryptococcal cell-mediated immune response. Infect Immun 1993; 61: 4750–4759.

Murphy JW, Hidore MR, Wong SC. Direct interactions of human lymphocytes with the yeast-like organism, *Cryptococcus neoformans*. J Clin Invest 1993; 91: 1553–1566.

Nabavi N, Murphy JW. In vitro binding of natural killer cells to *Cryptococcus neoformans* targets. Infect Immun 1985; 50: 50–57.

Nakamura LT, Wu-Hsieh BA, Howard DH. Recombinant murine gamma interferon stimulates macrophages of the RAW cell line to inhibit intracellular growth of *Histoplasma capsulatum*. Infect Immun 1994; 62: 680–684.

Negroni P. Estudio del primer caso argentino de cromomicosis, *Fonsecaea* (Neg.) *pedrosoi* (Brump) 1921. Dept Nat de Higiene Rev Inst Bacteriol 1936; 7: 419–426.

Neill JM, Sugg JY, McCauley DW. Serologically reactive material in spinal fluid, blood, and urine from a human case of cryptococcosis (torulosis). Proc Soc Exp Biol Med 1951; 77: 775–778.

Ness MJ, Vaughan WP, Woods GL. *Candida* antigen latex test for detection of invasive candidiasis in immunocompromised patients. J Infect Dis 1989; 159: 495–502.

Newberry WM, Chandler JW, Chin TD, Kirkpatrick CH. Immunology of the mycoses. I. Depressed lymphocyte transformation in chronic histoplasmosis. J Immunol 1968; 100: 436–443.

NIAID-MSG (National Institute of Allergy and Infectious Diseases Mycoses Study Group). Treatment of blastomycosis and histoplasmosis with ketoconazole. Ann Intern Med 1985; 103: 861–782.

Nichols EH. The relation of blastomycetes to cancer. J Med Res 1902; 7: 313–381.

Nickerson WJ. Reduction of inorganic substances by yeasts. I. Extracellular reduction of sulfite by species of *Candida*. J Infect Dis 1953; 93: 43–56.

Nickerson WJ, Mankowski Z. Role of nutrition in the maintenance of the yeast-shape in *Candida*. Am J Botany 1953; 40: 584–592.

Nickerson WJ. An enzymatic locus participating in cellular division of a yeast. J Gen Physiol 1954; 37: 483–494.

Olaiya AF, Sogin SJ. Ploidy determination of *Candida albicans*. J Bacteriol 1979; 140: 1043–1049.

Ophüls W, Moffitt HC. A new pathogenic mould. Preliminary report. Philadelphia Med J 1900; 5: 1471–1472.

Ophüls W. Further observations on a pathogenic mould formerly described as a protozoon (*Coccidioides immitis, Coccidioides pyogenes*). J Exp Med 1905; 6: 443–486.

Ouchterlony O. Antigen-antibody reactions in gels. Acta Pathol Microbiol Scand 1949; 26: 507–515.

Oxford AE, Raistrick H, Simonart P. Studies in the biochemistry of microorganisms. 60. Griseofulvin, a metabolic product of *Penicillium griseofulvum* Dierck. Biochem J 1939; 33: 240–248.

Palmer CE. Nontuberculous pulmonary calcification and sensitivity to histoplasmin. Public Health Reports Washington 1945; 60: 513–520.

Pappagianis D, Miller RL, Smith CE, Kobayashi GS. Response of monkeys to respiratory challenge following subcutaneous inoculation with *Coccidioides immitis*. Am Rev Respir Dis 1960; 82: 244–250.

Pappagianis D and the Valley Fever Vaccine Study Group. Evaluation of the protective efficacy of the killed *Coccidioides immitis* spherule vaccine in humans. Am Rev Respir Dis 1993; 148: 656–660.

Pappas PG, Threlkeld MG, Bedsole GD, Cleveland KO, Gelfand MS, Dismukes WE. Blastomycosis in immunocompromised patients. Medicine 1993; 72: 311–325.

Paltauf A. Mycosis mucorina: Ein Beitrag zur Kenntnis der menshlichen Fadenpilzer-Krankungen. Virchows Arch Pathol Anat 1885; 102: 543.

Pedroso A, Gomes JM. Four cases of dermatitis verrucosa produced by *Phialophora verrucosa*. Ann Paulistas Med Cir 1920; 9: 53.

Perfect JR, Magee BB, Magee PT. Separation of chromosomes of *Cryptococcus neoformans* by pulsed field gel electrophoresis. Infect Immun 1989; 57: 2624–2627.

Petersen RH. Lexemuel Ray Hesler. February 20, 1888–November 20, 1977. Mycologia 1978; 70: 757–765.

Pfaller MA, Rinaldi MG, Galgiani JN et al. Collaborative investigation of variables in susceptibility testing of yeasts. Antimicrob Agents Chemother1990; 34: 1648–1654.

Pfaller MA, Cabezudo I, Buschelman B et al. Value of the hybritech ICON *Candida* assay in the diagnosis of invasive candidiasis in high-risk patients. Diagn Microbiol Infect Dis 1993; 16: 53–60.

Phaff HJ. My life with yeasts. Ann Rev Microbiol 1986; 40: 1–28.

Pine L, Peacock CL. Studies on the growth of *Histoplasma capsulatum*. IV. Factors influencing conversion of the mycelial phase to the yeast phase. J Bacteriol 1958; 75: 167–174.

Pine L, Webster RE. Conversion in strains of *Histoplasma capsulatum*. J Bacteriol 1962; 83: 149–157.

Posada A. Un nuevo caso de micosis fungoidea con psorospermias. Anales de Circulo Medico Argentino 1892; 5: 585–597.

Powderly WG, Saag MS, Cloud GA et al. A controlled trial of fluconazole or amphotericin B to prevent relapse of cryptococcal meningitis in patients with the acquired immunodeficiency syndrome. N Engl J Med 1992; 326: 793–798.

Rapp F. The Friend legacy: From mouse to man. Ann NY Acad Sci 1989; 567: 349–353.

Rappaport BZ, Kaplan B. Generalized *Torula* mycosis. Arch Pathol Lab Med 1926; 1: 720–741.

Reiss E, Stone SH, Hasenclever HF. Serological and cellular immune activity of peptidoglucomannan fractions of *Candida albicans* cell walls. Infect Immun 1974; 9: 881–890.

Reiss E, Lehmann PF. The nature of the galoctomannan antigen of *Aspergillus fumigatus*. Infect Immun 1979; 25: 357–365.

Research Corporation. Brown-Hazen grants of $341,000 approved for biological sciences: New emphasis placed on medical mycology. Tucson, AZ: Spring Q Rep 1971: 7–8.

Research Corporation. Brown–Hazen program grants $157,000. Tucson, AZ. Summer Q Rep 1975:5.

Research Corporation. Gilbert Dalldorf and the Dalldorf Fellowship in Medical Mycology. Tucson, AZ, 1990. No author cited.

Research Corporation News. Two women scientists elected to the inventors hall of fame. Tucson, AZ, 1994. No author cited.

Resnick S, Pappagianis D, McKerrow JH. Proteinase production by the parasitic cycle of the pathogenic fungus *Coccidioides immitis*. Infect Immun 1987; 55: 2807–2815.

Rex JH, Pfaller MA, Galgiani JN et al. Development of interpretive breakpoints for antifungal susceptibility testing: Conceptual framework and analysis of in vitro-in vivo correlation data for fluconazole, itraconazole, and *Candida* infections. Clin Infect Dis 1997; 24: 235–247.

Rifkind D, Marchioro TL, Schneck SA, Hill RB. Systemic fungal infections complicating renal transplantation and immunosuppressive therapy. Am J Med 1967; 43: 28–38.

Riley WA, Watson CJ. Histoplasmosis of Darling with report of a case originating in Minnesota. Am J Trop Med 1926; 6: 271–282.

Rippon JW. Elastase: Production by ringworm fungi. Science 1967; 157: 947.

Rippon JW. Extracellular collagenase from *Trichophyton schoenleinii*. J Bacteriol 1968; 95: 43–46.

Rippon JW, Garber ED. Dermatophyte pathogenicity as a function of mating type and associated enzymes. J Invest Dermatol 1969; 53: 445–448.

Rixford E, Gilchrist TC. Two cases of protozoan (coccidioidal) infection of the skin and other organs. Johns Hopkins Hosp Rep 1896; 1: 209–268.

Robbins WJ, Ma R. Growth factors for *Trichophyton mentagrophytes*. Am J Botany 1945; 32: 509–523.

Roberts GD, Horstmeier C, Hall M, Washington JA. Recovery of yeast from vented blood culture bottles. J Clin Microbiol 1975; 2: 18–20.

Roberts GD, Wang HS, Hollick GE. Evaluation of the API 20C microtube system for the identification of clinically important yeasts. J Clin Microbiol 1976; 3: 302–305.

Robin C. Histoire naturelle des vegétaux parasites qui croissent sur l'homme et sur les animaux vivants. Paris: J.B. Bailliere, 1853.

Robinson JH. New History. New York: MacMillan, 1912.

Rockefeller Foundation. Unpublished letter to Dr. Butler. Medical Mycological Society of the Americas Archives, 1929.

Rogers DP. LD deSchweinitz and early American mycology. Mycologia 1977a; 69: 223–245.

Rogers DP. A Brief History of Mycology in North America. Second International Mycological Congress. Amherst, MA: Hamilton I. Newell, Inc, 1977b.

Rogerson CT. Fred Jay Seaver (1877–1970). Mycologia 1973; 65: 721–724.

Roilides E, Ühlig K, Venzon D, Pizzo PA, Walsh TJ. Prevention of corticosteroid-induced suppression of human polymorphonuclear leukocyte-induced damage of *Aspergillus fumigatus* hyphae by granulocyte colony-stimulating factor and gamma interferon. Infect Immun 1993; 61: 4870–4877.

Roilides E, Holmes A, Blake C, Pizzo PA, Walsh TJ. Impairment of neutrophil antifungal activity against hyphae of *Aspergillus fumigatus* in children infected with human immunodeficiency virus. J Infect Dis 1993; 16: 905–911.

Rosenthal T. David Gruby (1910–1898). Ann Med Hist 1932; 4: 339–346.

Rüchel R. Properties of a purified proteinase from the yeast *Candida albicans*. Biochem Bio Acta 1981; 659: 99–113.

Rudolph M. Ueber die brasilianisele figueira. Arch für Schiffs Tropeh-Hyg 1914; 18: 498.

Saag MS, Powderly WG, Cloud GA et al. Comparison of amphotericin B with fluconazole in the treatment of acute AIDS-associated cryptococcal meningitis. N Engl J Med 1992; 326: 83–89.

Sabouraud R. Les teignes. Paris: Mason et Cie, 1910.

Saccardo PA. Sylloge fungorum vol. 4. Published by the author, Pavia, Italy, 1886.

Saccardo PA. Sylloge fungorum vol. 18. Published by the author, Pavia, Italy, 1906.

Salkin IF. Further simplification of the *Guizotia abyssinica* seed medium for identification of *Cryptococcus neoformans* and *Cryptococcus bacillispora*. Can J Microbiol 1979; 25: 1116–1118.

Salvin SB. Cysteine and related compounds in the growth of the yeast-like phase of *Histoplasma capsulatum*. J Infect Dis 1949; 84: 275–283.

Salvin SB. Endotoxin in pathogenic fungi. J Immunol 1952; 69: 89–99.

Sanfelice F. Contributo alla morfologia e biologia dei blastomiceti che si sviluppano nei succhi di alcuni frutti. Ann Igiene 1894; 5: 239–262.

Sarachek A, Rhoads DD, Schwarzhoff RH. Hybridization of *Candida albicans* through fusion of protoplasts. Arch Microbiol 1981; 129: 1–8.

Schaefer JC, Yu B, Armstrong D. An *Aspergillus* immunodiffusion test in the early diagnosis of aspergillosis in adult leukemia patients. Am Rev Respir Dis 1976; 113: 325–329.

Schenck BR. On refractory subcutaneous abscesses caused by a fungus possibly related to the sporotricha. Johns Hopkins Hosp Bull 1898; 9: 286–290.

Scherer S, Davis RW. Replacement of chromosome segments with altered DNA sequences constructed in vitro. Proc Nat Acad Sci 1979; 76: 4951–4955.

Scherer S, Stevens DA. Application of DNA typing methods to epidemiology and taxonomy of *Candida* species. J Clin Microbiol 1987; 25: 675–679.

Scherer S, Stevens DA. A *Candida albicans* dispersed, repeated gene family and its epidemiology applications. Proc Nat Acad Sci USA 1988; 85: 1452–1456.

Schoenbach EB, Miller JM, Ginsberg M, Long PH. Systemic blastomycosis treated with stilbamidine. J Am Med Assoc 1951; 146, 1317–1318.

Schoenlein JL. Zur pathogene der impetigines. Arch Anat Physiol Wiss Med (Müller's) 1839; 82, Taf.

Schwartz DC, Cantor CR. Separation of yeast chromosome-sized DNAs by pulsed field gradient gel electrophoresis. Cell 1984; 37: 67–75.

Schwarz J, Baum GL. Blastomycosis. Am J Clin Pathol 1951; 21: 999–1029.

Schwarz J, Baum GL Pioneers in the discovery of deep fungus diseases. Mycopathol Mycol Applicata 1964; 25: 73–81.

Seeber GR. Un nuevo esporozuario parasito del hombre. Dos casos encontrados en polipos nasales. Tesis, Universidad Nacional de Buenos Aires, 1900.

Seabury JH, Dascomb HE. Experience with amphotericin B for the treatment of systemic mycoses. Arch Intern Med 1958; 102: 960–976.

Seeliger HP, Seefried L. Aldo Castellani – An appraisal of his life and oeuvre. Mycoses 1989; 32: 391.

Segal E, Berg RA, Pizzo PA, Bennett JE. Detection of *Candida* antigen in sera of patients with candidiasis by an enzyme-linked immunosorbent assay-inhibition technique. J Clin Microbiol 1979; 10: 116–118.

Selsted ME, Brown DM, DeLange RJ, Harwig SL, Lehrer RI. Primary structures of six antimicrobial peptides of rabbit peritoneal neutrophils. J Biol Chem 1985; 260: 4579–4584.

Sequeira L. William H. Weston (1890–1978): Tribute and remembrance. Annu Rev Phytopathol 1993; 31: 43–52.

Shadomy HJ. Clamp connections in two strains of *Cryptococcus neoformans*. Spectrum monograph series, Arts Science, Georgia State University, Atlanta 1970; 1: 67–72.

Shadomy S. In vitro studies with 5-fluorocytosine. Appl Microbiol 1969; 17: 871–877.

Shaffer MF. History of Medical Mycology at Tulane Medical School. Unpublished manuscript, Tulane University, New Orleans, 1985.

Sharkey-Mathis PK, Kauffman CA, Graybill JR et al. Treatment of sporotrichosis with itraconazole. Am J Med 1993; 95: 279–285.

Shear CL. Life history of an undescribed ascomycete isolated from a granular mycetoma of man. Mycologia 1922; 14: 239–243 .

Sidransky H, Friedman L. The effect of cortisone and antibiotic agents on experimental pulmonary aspergillosis. Am J Pathol 1959; 35: 169–183.

Silva M, Benham RW. Nutritional studies of the dermatophytes with special reference to *Trichophyton megnini* Blanchard 1896 and *Trichophyton gallinae* (Megnin 1881) comb.nov. J Invest Dermatol 1952; 18: 453–472.

Silva M, Benham RW. Nutritional studies of the dermatophytes with special reference to the red-pigment producing-varieties of *Trichophyton mentagrophytes*. J Invest Dermatol 1954; 22: 285–294.

Silva M, Kesten BM, Benham RW. *Trichophyton rubrum* infections: A clinical, mycologic, and experimental study. J Invest Dermatol 1955; 25: 311–328.

Silva ME, Hazen EL. Rhoda Williams Benham. Mycologia 1957; 49: 596–603.

Silva-Hutner ME. Brief History of the Mycology Laboratory, Columbia–Presbyterian Medical Center. Unpublished manuscript from Medical Mycological Society of the Americas Archives, 1975.

Slutsky B, Buffo J, Soll DR. High-frequency switching of colony morphology in *Candida albicans*. Science 1985; 230: 666–669.

Smit P, Heniger J. Antoni van Leeuwenhoek (1632–1723) and the discovery of bacteria. Antonie van Leeuwenhoek 1975; 41: 219–28.

Smith LW, Sano ME. Moniliasis with meningeal involvement. J Infect Dis 1933; 53: 187–196.

Smith CE. Epidemiology of acute coccidioidomycosis with erythema nodosum. Am J Public Health, 1940; 30: 600–611.

Smith CE, Beard RR, Rosenberger HG, Whiting EG. Effect of season and dust control on coccidioidomycosis. J Am Med Assoc 1946; 132: 833–838.

Smith CE, Whiting EG, Baker EE, Rosenberger HG, Beard RR, Saito MT. The use of coccidioidin. Am Rev Tuberc 1948; 57: 330–360.

Smith CE, Saito MT, Beard RR, Kepp RM, Clark RW, Eddie BU. Serological tests in the diagnosis and prognosis of coccidioidomycosis. Am J Hygiene 1950; 52: 1–21.

Snapper I, McVay LV. The treatment of North American blastomycosis with 2-hydroxystilbamidine. Am J Med 1953; 15: 603–623.

Sokol-Anderson ML, Brajtburg J, Medoff G. Amphotericin B-induced oxidative damage and killing of *Candida albicans*. J Infect Dis 1986; 154: 76–83.

Soll DR, Langtimm CJ, McDowell J, Hicks J, Galask R. High frequency switching in *Candida* strains isolated from vaginitis patients. J Clin Microbiol 1987; 25: 1611–1622.

Southern EM. Detection of specific sequences among DNA fragments separated by gel electrophoresis. J Molec Biol 1975; 98: 503–518.

Spitzer ED, Keath EJ, Travis SJ, Painter AA, Kobayashi GS, Medoff G. Temperature-sensitive variants of *Histoplasma capsulatum* isolated from patients with acquired immunodeficiency syndrome. J Infect Dis 1990; 162: 258–261.

Spitzer ED, Spitzer SG. Use of a dispersed repetitive DNA element to distinguish clinical isolates of *Cryptococcus neoformans*. J Clin Microbiol 1992; 30: 1094–1097b.

Srikantha T, Soll DR. A white specific gene in the white-opaque switching system of *Candida albicans*. Gene 1993; 131: 53–60.

Staib F. *Cryptococcus neoformans* und *Guizotia abyssinica* (syn. *G. oleifera* D.C.). Z Hyg Infektionskr 1962; 148: 456–475.

Staib F. Serum-proteins as nitrogen source for yeast-like fungi. Sabouraudia 1965; 4: 187–193.

Stamm AM, Diasio RB, Dismukes WE et al. Toxicity of amphotericin B plus flucytosine in 194 patients with cryptococcal meningitis. Am J Med 1987; 83: 236–242.

Standard PG, Kaufman L. Specific immunological test for the rapid identification of members of the genus *Histoplasma*. J Clin Microbiol 1976; 3: 191–199.

Standard PG, Kaufman L. Immunological procedure for the rapid and specific identification of *Coccidioides immitis* cultures. J Clin Microbiol 1977; 5: 149–153.

Steinberg BA, Jambor WP, Suydam LO. Amphotericins A and B: Two new antifungal antibiotics possessing high activity against deep-seated and superficial mycoses. Antibiot Annu 1955–1956, 574–578 .

Stenderup A, Bak AL. Deoxyribonucleic acid base composition of some species within the genus *Candida*. J Gen Microbiol 1968; 52: 231–236.

Stern JJ, Hartman BJ, Sharkey P et al. Oral fluconazole therapy for patients with acquired immunodeficiency syndrome and cryptococcosis: experience with 22 patients. Am J Med 1988; 85: 477–480.

Stevens P, Huang S, Young LS, Berdischewsky M. Detection of *Candida* antigenemia in human invasive candidiasis by a new solid phase radioimmunoassay. Infection 1980; 8: S334–S338.

Stevens DA, Greene SI, Lang OS. Thrush can be prevented in patients with acquired immunodeficiency syndrome and the acquired immunodeficiency syndrome-related complex. Arch Intern Med 1991; 151: 2458–2464.

Stewart RA, Meyer KF. Isolation of *Coccidioides immitis* (Stiles) from the soil. Proc Soc Exp Biol Med 1932; 29: 937–938.

Stockdale PM. *The Microsporum gypseum* complex (*Nannizzia incurvata* Stockd, *N. gypsea* (Nann.) comb. nov. *N. fulva* sp. nov.). Sabouraudia 1963; 3: 114–126.

Stockman L, Clark KA, Hunt JM, Roberts GD. Evaluation of commercially available acridinium ester-labeled chemiluminescent DNA probes for culture identification of *Blastomyces dermatitidis*, *Coccidioides immitis*, *Cryptococcus neofomans*, and *Histoplasma capsulatum*. J Clin Microbiol 1993; 31: 845–850.

Stoddard JL, Cutler EC. *Torula* infection in man. Rockefeller Institute for Medical Research, Monograph, No. 6, 1–98, 1916.

Storck R. Nucleotide composition of nucleic acids of fungi: II. Deoxyribonucleic acids. J Bacteriol 1966; 91: 227–230.

Subramanian CV. Hyphomycetes: Taxonomy and Biology. London: Academic Press, 1983.

Sugar AM, Saunders C. Oral fluconazole as suppressive therapy of disseminated cryptococcosis in patients with acquired immunodeficiency syndrome. Am J Med 1988; 85: 481–489.

Sugar AM, Picard M. Macrophage-and oxidant–mediated inhibition of the ability of live *Blastomyces dermatitidis* conidia to transform to the pathogenic yeast phase: Implications for the pathogenesis of dimorphic fungal infections. J Infect Dis 1991; 163: 371–375.

Sullivan PA, Wright RJ. Two genes for secreted aspartate proteinase in *Candida albicans*. Abstr. 8th International Symposium of Yeast, G-10, 1992: 203.

Swartz JH, Conant NF. Direct microscopic examination of the skin. A method for the determination of the presence of fungi. Arch Dermatol Syphilol 1936; 33: 291–305.

Sweany HC. *Histoplasmosis*. Springfield, Illinois: Charles C. Thomas, 1960.

Switchenko AC, Miyada CG, Goodman TC et al. An automated enzymatic method for measurement of D-arabinitol, a metabolite of pathogenic *Candida* species. J Clin Microbiol 1984; 32: 92–97.

Syverson RE, Buckley HR, Campbell CC. Cytoplasmic antigens unique to the mycelial or yeast phase of *Candida albicans*. Infect Immun 1975; 12: 1184–1188.

Szaniszlo PJ, Hsieh PH, Marlowe JD. Induction and ultrastructure of the multi-cellular (sclerotic) morphology in *Phialophora dermatitidis*. Mycologia 1976; 48: 117–130.

Szaniszlo P. John Nathaniel Couch, 1896–1986. National Academy of Sciences, Biographical Memoirs, 62, 39–64. Washington, DC: The National Academy Press, 1993.

Taplin D, Zaias N, Rebell G, Blank H. Isolation and recognition of dermatophytes on a new media (DTM). Arch Dermatology 1969; 99: 203–209.

Tassel D, Madoff MA. Treatment of *Candida* sepsis and *Cryptococcus* meningitis with 5-fluorocytosine: A new antifungal agent. J the Am Med Assoc 1968; 206: 830–832.

Tavares II. Lee Bonar (1891–1977). Mycologia 1979; 71: 681–687.

Titsworth E, Grunberg E. Chemotherapeutic activity of 5-fluorocytosine and amphotericin B against *Candida albicans* in mice. Antimicrob Agents Chemother 1973; 4: 306–308.

Todd RL. Studies on yeast-like organisms isolated from the mouths and throats of normal persons. Am J Hygiene 1937; 25: 212–220.

Tompkins L, Plorde JJ, Falkow S. Molecular analysis of R-factors from multiresistant nosocomial isolates. J Infect Dis 1980; 141: 625–636.

Torack RM. Fungus infections associated with antibiotic and steroid therapy. Am J Med 1957; 22: 872–882.

Turner TB. History of medical education at Johns Hopkins. Johns Hopkins Med J 1976; 139: 27–36.

Utz JP, Louria DB, Feder N, Emmons CW, McCullough NB. A report of clinical studies on the use of Amphotericin B in patients with systemic fungal diseases. Antibiot Annu 1957–1958, 65–70.

Utz JP, Treger A, McCullough NB, Emmons CW. Amphotericin B: Intravenous use in 21 patients with systemic fungal diseases. Antibiot Annu 1958–1959, 628–634.

Utz JP, Tynes BS, Shadomy HJ, Duma RJ, Kannan MM, Mason KN. 5-fluorocytosine in human cryptococcosis. Antimicrob Agents Chemother 1968; 8: 344–349.

Utz JP, Garriques IL, Sande MA et al. Therapy of cryptococcosis with combination of flucytosine and amphotericin B. J Infect Dis 1975; 132: 368–373.

Van Der Horst CM, Saag MS, Cloud GA et al. Treatment of acute cryptococcal meningitis associated with the acquired immunodeficiency syndrome. N Engl J Med. 1997; 337: 15–21.

Van Pernis PA, Benson ME, Holinger PH. Specific cutaneous reactions with histoplasmosis. J the Am Med Assoc 1941; 117: 436–437.

Vasquez JA, Beckley A, Donabedian S, Sobel JD, Zervos MJ. Comparison of restriction enzyme analysis versus pulsed-field gradient electrophoresis as a typing system for *Torulopsis glabrata* and *Candida* species other than *C. albicans*. J Clin Microbiol 1993; 31: 2021–2030.

Vogel RA, Padula JF. Indirect staining reaction with fluorescent antibody for detection of antibodies to pathogenic fungi. Proc Soc Exp Biol Med 1958; 98: 135–139.

Vogel RA. The indirect fluorescent antibody test for the detection of antibody in human cryptococcal disease. J Infect Dis 1966; 116: 573–580.

Vollrath D, Davis RW. Resolution of DNA molecules greater than 5 megabases by countour-clamped homogeneous electric fields. Nucleic Acids Res 1987; 15: 7865–7876.

Vuillemin P. Le Blastomycates pathogenes. Rev Gen Sci 1901; 12: 732–751.

Wadsworth E, Prasad SC, Calderone R. Analysis of mannoproteins from blastoconidia and hyphae of *Candida albicans* with a common epitope recognized by anti-complement receptor type 2 antibodies. Infect Immun 1993; 61: 4675–4681.

Waldorf AR, Levitz SM, Diamond RD. In vivo bronchoalveolar macrophage defense against *Rhizopus oryzae* and *Aspergillus fumigatus*. J Infect Dis 1984; 150: 752–760.

Walker JW, Montgomery FH. Further report of a previously recorded case of blastomycosis of the skin; systemic infection with blastomyces; death; autopsy. J Am Med Assoc 1902; 38: 867–871.

Walsh LR, Poupard JA. Lydia Rabinowitsch, Ph.D, and the emergence of clinical pathology in late 19th-century America. Arch Pathol Lab Med 1989; 113: 1303–1308.

Walsh TJ, Lee J, Aoki S et al. Experimental basis for use of fluconazole for preventive or early treatment of disseminated candidiasis in granulocytopenic hosts. Rev Infect Dis 1990; 12: S307–S317.

Walsh TJ, Hathorn JW, Sobel JD et al. Detection of circulating *Candida* enolase by immunoassay in patients with cancer and invasive candidiasis. N Engl J Med 1991; 324: 1026–1031.

Walsh TJ, Merz WG, Lee JW et al. Diagnosis and therapeutic monitoring of invasive candidiasis by rapid enzymatic detection of serum D-arabinitol. Am J Med 1995; 99: 164–172.

Warren KS. Tropical medicine or tropical health: The Heath Clark Lectures, 1988. Rev Infect Dis 1990; 12: 142–156.

Watson DW. Who speaks for microbiology? Bacteriol Rev 1969; 33(3): 383–389.

Weidman FD, Spring D. Comparison of ringworm culture ingredients: II and III. Arch Dermatol Syphilol 1928; 18: 829–851.

Weiner MH, Yount WJ. Mannan antigenemia in the diagnosis of invasive *Candida* infections. J Clin Invest 1976; 58: 1045–1053.

Weiner MH, Coats-Stephen M. Immunodiagnosis of systemic aspergillosis. I. Antigenemia detected by radioimmunoassay in experimental infection. J Lab Clin Med 1979; 93: 111–119.

Weiner MH. Antigenemia detected by radioimmunoassay in systemic aspergillosis. Ann Intern Med 1980; 92: 793–796.

Weiner MH. Antigenemia detected in human coccidioidomycosis. J Clin Microbiol 1983; 18: 136–142.

Weitzman I. Incompatibility in the *Microsporum gypseum* complex. Mycologia 1964a; 56: 425–435.

Weitzman I. Variation in *Microsporum gypseum*. I. A genetic study of pleomorphism. Sabouraudia 1964b; 3: 195–204.

Weitzman I, McGinnis MR, Padhye AA, Ajello L. The genus *Arthroderma* and its later synonym *Nannizzia*. Mycotaxon 1986; 25: 505–518.

Wheat LJ, Kohler RB, Tewari RP. Diagnosis of disseminated histoplasmosis by detection of *Histoplasma capsulatum* antigen in serum and urine specimens. N Engl J Med 1986; 314: 83–88.

Wheat JL, Connolly-Stringfield P, Blair R, Connolly K, Garringer T, Katz BP. Histoplasmosis relapse in patients with AIDS: Detection using *Histoplasma capsulatum* variety *capsulatum* antigen levels. Ann Intern Med 1991; 115: 936–941.

Wheat J, Hafner R, Wulfsohn M et al. Prevention of relapse of histoplasmosis with itraconazole in patients with the acquired immunodeficiency syndrome. Ann Intern Med 1993; 118: 610–616.

Whelan WL, Partridge RM, Magee PT. Heterozygosity and segregation in *Candida albicans*. Mol Gen Genet 1980; 180: 107–113.

Whitaker RH. New concepts of kingdoms of organisms. Science 1969; 163: 150–160.

White CJ. Ringworm as it exists in Boston. J Cutan Genito-Urin Dis 1899; 17: 1–17.

Wickerham LJ, Burton KA. Carbon assimilation tests for the classification of yeasts. J Bacteriol 1948; 56: 363–371.

Wickes BL, Golin JE, Kwon-Chung KJ. Chromosomal rearrangement in *Candida stellatoidea* results in a positive effect on phenotype. Infect Immun 1991; 59: 1762–1771.

Wickes BL, Staudinger J, Magee BB, Kwon-Chung KJ, Magee PT, Scherer S. Physical and genetic mapping of *Candida albicans*: Several genes previously assigned to chromosome 1 map to chromosome R, the rDNA-containing linkage group. Infect Immun 1991; 59: 2480–2484.

Williams JGK, Kubelik AR, Livak KJ, Rafalski JA, Tingey SV. DNA polymorphisms amplified by arbitrary primers are useful as genetic markers. Nucleic Acids Res 1990; 18: 6531–6535.

Wilson CM. William H. Weston. Mycologia 1979; 71: 1103–1106.

Wilson PW. Training a microbiologist. Annu Rev Microbiol 1972; 26: 1–22.

Wilson JW, Smith CE, Plunkett OA. Primary cutaneous coccidioidomycosis: The criteria for diagnosis and a report of a case. California Med 1953; 79: 233–239.

Wingard JR, Merz WG, Saral R. *Candida tropicalis*: A major pathogen in immunocompromised patients. Ann Intern Med 1979; 91: 539–543.

Wingard JR, Merz WG, Rinaldi MG, Johnson TR, Karp JE, Saral R. Increase in *Candida krusei* infection among patients with bone marrow transplantation and nutropenia treated prophylactically with fluconazole. N Engl J Med 1991; 325: 1274–1277.

Winn WA. Coccidioidomycosis and amphotericin B. Med Clin N Am 1963; 47: 1131–1148.

Wong B, Brauer KL. Enantioselective measurement of fungal D-arabinitol in the sera of normal adults and patients with candidiasis. J Clin Microbiol 1988; 26: 1670–1674.

Wong B, Murray JS, Castellanos M, Croen KD. D-arabitol metabolism in *Candida albicans*: Studies of the biosynthetic pathway and the gene that encodes NAD-dependent D-arabitol dehydrogenase. J Bacteriol 1993; 175: 6314–6320.

Woods JP, Kersulyte D, Goldman WE, Berg DE. Fast DNA isolation from *Histoplasma capsulatum*: Methodology for arbitrary primer polymerase chain reaction-based epidemiological and clinical studies. J Clin Microbiol 1993; 31: 463–464.

Woodward TE. The golden era of microbiology: People and events of the 1880s. MD Med J 1989; 38(4): 323–328.

Woolley DW. Some biological effects produced by benzimidazole and their reversal by purines. J Biol Chem 1944; 152: 225–232.

Wright JH. A case of mycetoma (madura foot). J Exp Med 1898; 3: 421–433.

Wu-Hsieh B, Howard DH. Inhibition of growth of *Histoplasma capsulatum* by lymphokine-stimulated macrophages. J Immunol 1984; 132: 2593–2597.

Young RC, Bennett JE, Vogel CL, Carbone PP, DeVita VT. Aspergillosis. The spectrum of the disease in 98 patients. Medicine 1970; 49: 147–173.

Yuan L, Cole GT. Isolation and characterization of an extracellular proteinase of *Coccidioides immitis*. Infect Immun 1987; 55: 1970–1978.

Zakon SJ, Benedek T. David Gruby and the centenary of medical mycology (1841–1941). Bull Hist Med 1944; 16: 155–168.

Zeidberg LD, Ajello L, Dillon A, Runyon LC. Isolation of *Histoplasma capsulatum* from soil. Am J Public Health 1952; 42: 930–935.

Ziegler-Bohme H, Gemeinhardt H. 150 Jahre medizinische Mykoloie seit Johann Lucas Schonlein [150 years of medical mycology since Johann Schonlein]. Geschichte der Dermatologie, 1990; 176: 623–631.

Zimmerman LE. *Candida* and *Aspergillus* endocarditis. Arch Pathol 1950; 50: 591–605.

Zimmerman LE, Rappaport H. Occurrence of cryptococcosis in patients with malignant disease of reticuloendothelial system. Am J Clin Pathol 1954; 24: 1050–1072.

Zopf W. Die Pilz im Morphologischer, Physiologischer, Biologischer Beziehung. Breslau, Trewendt, 1890.

Appendix A: Questionnaire analysis summary

An open-ended questionnaire was developed to survey a diverse sample of medical scientists, including retirees in the field of medical mycology. The purpose of this survey was to investigate in a historical context the development of medical mycology in the United States. The questions were structured to solicit the thoughts of medical mycologists regarding significant contributors and historically important educational and scientific events in the development of medical mycology in the U.S. since 1894. Ideas were solicited on the current state of the discipline and what its direction may be in the near future.

Analysis

The 109 final questionnaires received represent a population sample of 35% of the 320 individual mycologists contacted. A list of the 109 respondents also is included in Appendix E. Their responses have proven to be invaluable in giving clarity, providing insight and expanding the depth of this research. The results of these responses are described in the following discussion.

The first question dealt with the amount of time the respondent devoted to medical mycology (Table 1). The sample population was almost evenly (16.2 to 24.7%) distributed in the five time categories listed in the questionnaire, so they were consolidated into three categories: 1 to 25%, 26 to 50% and 51 to 100%. Nearly half of the respondents spend more that 50% of their time in medical mycology and another 20% spend more than 25%.

The data summarized in Table 2 revealed that of their various activities, 90% of the respondents spend some time on medical mycology *research*. A significant result was that 41% spend more than 25% of their time in such *research*, the next closest category being *informal teaching* at 10.5%. After *research*, the number of scientists spending some time in other medical mycology activities gradually declines with the second activity being *formal teaching* (84.8%, 89 individuals). But the vast majority of this group (78 or 74.3%) spend 25% or less of their time here. The next three activities are closely

grouped, whereby the respondents spend 74.3%, 70.5%, and 68.6% of their time doing consultation, writing or informal teaching, respectively.

Question three asked, *What was your definition of a medical mycologist?* The definitional responses to this question were diverse and centered on the amount of time devoted to the field. In summary, the respondents stated that medical mycologists are individuals who are devoted to the study of pathogenic fungi in humans. Their primary activities are research, teaching/training, and clinical laboratory work.

Question four asked, *What credentials/experiences are needed to be considered a medical mycologist?* Formal education was the top response to this open-ended question as 55.6% stated that at least a B.S. degree was needed in combination with experience and/or medical mycology training.

The focus of question five was the *single most important contribution* to *the development of medical mycology in the U.S.* Table 3 lists a range of contributions as noted by 86 (79.9%) of the 109 respondents. Nineteen (19) individuals identified the work of Norman Conant as the most important contribution (22.1%). Conant's book in particular was noted because it was one of the first efforts to deal with all aspects of medical mycology. This early work demonstrated that fungal diseases were not as rare as generally thought within the medical community. His book gave direction to medical mycology, placed the discipline in formal academic settings, and gave laboratories the capacity to identify fungi quickly and accurately.

When asked, *What other major contributions have advanced the development of medical mycology in the US*, the top responses were the recognition of fungi as etiologic agents (30%) and training (and research) programs (30%). The importance of antifungal drugs followed closely at 24.7% (Table 4).

The responses to the question, *Who have been the major contributors to the discipline,* were most helpful in writing this paper, in that they provided a broad-based verification of the research conducted to trace the work of our pioneers. Also, the responses have clarified linkages between contributors and the contributions that they made to furthering training and research. Three pioneers in medical mycology were most frequently noted by the 98 respondents to this question: Chester Emmons (40.8%), Norman Conant (33.7%) and Libero Ajello (29.6%). As should be expected, a number of respondents focused on key figures during the last 30–40 years and were not as familiar with other early pioneers such as Rhoda Benham, Charles E. Smith, and Howard Larsh. The training and education contributions of important leaders in the field from 1920 to 1950 were validated by responses from medical mycologists who trained and worked with them.

When asked, *How have you contributed to the development of medical mycology,* two main areas dominated the answers of the 96 respondents: research and publications, especially of scientific papers, and teaching/training. Once again, because of multiple responses, there were 198 total contributions listed. The most frequently listed response category was research and the publication of their work, where 45.8% cited the publication of scientific papers. Closely following were the responses from 40.6% who noted that applied research was where they made contributions. Teaching/training was the second area overall that was most frequently noted, as teaching a medical mycology course (17.7%) and teaching physicians and/or medical technicians (14.6%) were the two dominant responses.

An important outcome of the questionnaire was information that led to the *creation of "Degree Trees" and "Mentor Trees".* The information compiled from these questions (defined as *"Training Trees"*) is presented in the Appendix B. The *"Tree"* results are organized by the University, Hospital, or Center where the respondents to this questionnaire worked and/or received training. In some instances, the *"Trees"* have been supplemented by information from curriculum vitas, research findings and/or personal interviews. By no means are the *"Trees"* inclusive of all degree students, professors, and mentors or mentees at a particular institution. All information is displayed by institution with major professors/trainers appearing in italics above each individual as appropriate and available. The mentor/mentee section in each *"Tree"* is listed as defined by the mentee, and in some cases there is overlap between major professor and mentor. The *"Trees"* are indicative of major centers of education, training, and research and have been helpful as a cross reference to the historical investigation conducted for this book. It is intended that the *"Trees"* will

continue to generate discussion and promote interest in defining and placing the many contributors to the field of medical mycology.

Since the *"Trees"* information is best understood in a display format, only some of the leading centers of formal education, training and research as well as informal training in medical mycology will be listed. Again, the primary basis of this information is the results of the questionnaire. Universities known and mentioned for educating and training graduates in medical mycology, among others were: Columbia, Duke, Harvard, Tulane, California at Los Angeles, California at Berkeley, Oklahoma, Kentucky, North Carolina, Michigan State, Montana State, Texas at Austin, and MCV/Virginia Commonwealth. Centers primarily noted for their work with postdoctoral students are the National Institutes of Health, the Centers for Disease Control, and the Universities of Alabama, Iowa, Texas at San Antonio, as well as Boston, Washington University, and Santa Clara.

The question, *Have your funding sources shifted from government sponsored grants to other sources,* provided mixed results. In regard to *research,* fifty-eight (58) individuals responded and fifty-one (51) gave no answer or said the question was not appropriate for them. A shift in funding was confirmed by 51.7% of these respondents, while 48.3% had no change in funding sources. The primary years when the shifts occurred were from 1972 to 1994, but 66.7% of these "yes" respondents said the shifts took place between 1980 and 1994.

Much less activity was evident in training grants as only 33% of all questionnaire respondents answered this question. Of those respondents only seven (7) indicated that a shift in funding sources had taken place. The primary years of the shifts were from 1982 to 1991. Of the respondents to this question, 80.6% indicated they had no shift in funding source.

When asked, *What other sources of funds support your current efforts in research,* a total of seventy-five (75) questionnaire respondents answered this question, or 68.8%. Of those affirming other sources of funds for *research,* 58.7% said their support came from industry and 65.9% of this group (29) receive their main support from the pharmaceutical industry. The remaining thirty-one individuals (41.3%) receive their support from a wide variety of sources including universities, foundations, government grants, private groups, etc. In regard to *training,* the majority (63.3%) of the questionnaire respondents did not answer this question. Of the 37.6% receiving training funds, no dominant source was mentioned, but the largest block of support came from industry (31.7%) while institutions fund 22% of this population.

Approximately half (51.4%) of the questionnaire respondents did not answer the question regarding *the effect of shifting of funding sources on their academic and/or research priorities.* A group of 33.9% respondents stated they had no change in priorities because of funding sources and 14.7% replied affirmatively. These respondents have either abandoned research they were doing or have turned to contracts and clinical applications, mostly with industry and other private sources. The main period of the shift in funds was from 1978 to 1993. Medical mycology research appears to be a limited field at this time as indicated by these responses. It is clear that the individuals conducting medical mycology training have had few new sources of funds and have either stopped their training work or support it as part of other activities.

There were 128 responses (96 individuals, 88.1%) to the question, *What is your perception of the current state of medical mycology in the U.S. (including education and training)?* Although the responses were quite diverse, by far the largest group (37) stated that there is a training crisis. Opinions were equally balanced on whether the current state of the discipline is either good or poor. Sixteen individuals described the discipline as "very healthy, expanding, at the apex of interest, poised on the edge of great progress", etc. In contrast, 17 others believe that medical mycology is "basically in a crisis, not strong, in extremis, endangered, tenuous", etc. Another eight respondents believe that the discipline is ether improving (3) or that there is room for improvement. Among the other responses were 13 with no opinion, 15 seeing a shift to molecular, and 22 with scattered, miscellaneous views such as, lack of financial support and visibility (9), inadequate communication channels or dialogue (3), scarcity of jobs (4), etc.

Of particular interest is the perception of a training crisis. The main reasons for this crisis were the lack of enough, as well as adequate, training programs, courses, and seminars at all levels. In general, respondents stated that medical mycology is under

emphasized in education and training as well as declining in the number and quality of diagnostic laboratories with trained personnel to staff them. They noted that while the field may have grown in importance and basic research is of high quality, few funds are available for applied research. These views coincide with those stating that there is a shift toward molecular biology basic research.

A total of 88 individuals responded to the question, *What changes do you think are needed for the advancement of medical mycology (including education and training)?* Among the respondents, 32 individuals saw the need for specific increases in medical mycology training at the undergraduate, graduate and postdoctoral levels, as well as in the medical schools and with more clinical orientation. Closely related to these views were another 32 respondents who focused on the need for more federal, state, and pharmaceutical funds for training and practical and basic research. When viewed as a whole, these 64 individuals make a clear statement that training, and the funds to support it, is a predominant need in medical mycology. The other 24 individuals indicated that additional changes in direction are needed such as better channels of communication, replacement of retirees, more basic research as needed, transference of funds from older to younger investigators, and more cooperative studies.

The last question was concerned with, *How should the changes be accomplished (including education and training)?* Of the 109 questionnaires, 73 responded while 36 either had no good ideas, did not know, or felt the question was very difficult. Among the 73 individuals, there were multiple responses that exceed the total number of respondents. A group of 40 responses targeted the need for more formal and informal training to accomplish the changes needed, with 25 of these individuals emphasizing more formal university training. Increased funding was seen as the strategic means to effect the changes listed in 25 responses. A strong majority of 19 in this group targeted the need for more NIH funding as a top priority. Another group of 17 responses focused on the need for improving public relations and promoting the discipline in a variety of other ways.

Table 1. Results of question no. 1 – professional time devoted to medical mycology (105 respondents)

	1–10%		11–25%	26–50%	51–75%		76–100%
No. of respondents	17		17	21	26		24
Total		34				50	
% of respondents	16.2		16.2	20.0	24.7		22.9
Total		32.4				47.6	

Total: 109

Table 2. Results of question no. 2 – professional time devoted to specific activities (105 respondents)

Rank order	Percent of the total	Less than 25% of time on	26–75% of time on
Research	90.5	41.0	41.0
Formal teaching	84.8	74.3	9.5
Consultation	74.3	63.8	7.6
Writing	70.5	60.0	9.5
Informal teaching	68.6	58.1	10.5
Administration	54.3	49.5	4.8
Clinical laboratory	52.4	40.0	8.6
Professional contributions	51.4	49.5	1.0
Patient care	38.1	29.5	8.6
Other	14.3	13.3	0

Total: 109

Table 3. Results of question no. 5 – single most important contribution (86 respondents, 79%)

Contribution (no. of respondents)	No. of answers (%)
Norman Conant	19 (22.1)
Book (14)	
Course (4)	
Book and course (1)	
Antifungal drugs	13 (15.1)
Immunosuppressed (8) and AIDS Pts (5)	13 (15.1)
Training	10 (11.6)
Programs (8) (e.g., Mayo Clinic)	
Books (2) (M. McGinnis', J. Rippon's)	
Institutions and groups	8 (9.3)
NIH (3) (C. Emmons)	
NIAID-MSG (2)	
NCCLS standardization of testing (1)	
CDC (1)	
MCV (1)	
Key figures	8 (9.3)
Sabouraud (5)	
Others (3)	
Recognition of fungi as etiologic agents	4 (4.7)
Molecular biology and genetics	3 (3.5)
Others	8 (9.3)
Research programs (1)	
Perfect stage of fungi (1)	
C. neoformans latex test (1)	
Benham's laboratory (1)	
Brown and Hazen Funds (1)	
WW II task force (1)	
Skin test development (1)	
Cumulative finding that human pathogens are only a subset of fungi (1)	
No answer	23 (21.1)
TOTAL	109

148

Table 4. Results of question no. 6 – other major contributions (89 respondents, 81.7%)

Contribution	No. of answers (%)
Recognition of fungi as etiologic agents	24 (30)
Training [research and training program(s)]	24 (30)
Antifungal drugs	22 (24.7)
Molecular tools/yeast genetics/molecular biology/gene description	14 (15.7)
NIH support/and medical mycology unit	14 (15.7)
HIV and immunosuppressed patients	13 (14.6)
Understanding of epidemiology	13 (14.6)
Individuals (including spokespersons)	12 (13.5)
Books	9 (10.1)
Diagnostic and rapid tests (non-serological)	9 (10.1)
Serological tests	8 (9)
Immunology research (host–pathogen interactions)	5 (5.6)
Commercial tests (Calcofluor and API 20C)	5 (5.6)
CDC medical mycology unit	4 (4.5)
Sexual state of pathogens	4 (4.5)
Clinical research and MSG	3 (3.4)
Societies	3 (3.4)
Ecological studies	3 (3.4)
McGinnis' taxonomical studies	2 (2.2)
Antifungal testing	2 (2.2)
Pfizer support/Brown and Hazen funding	2 (2.2)
The need to know/understand	2 (2.2)
Cooperative working groups	2 (2.2)
Pathogenecity studies	2 (2.2)
Understanding of complex culture conditions	2 (2.2)
Publications and meetings	2 (2.2)
Proficiency testing	2 (2.2)
No answer	20 (18.3)
TOTAL	109

Appendix B: Genealogic "Training Trees"
Listed by University or Medical Center

BOSTON UNIVERSITY

R Diamond, MD

Degree	Postdoc	Mentees
C Lyman, PhD	E Smail, MD (Dalldorf Fellow, 1983)	
	S Levitz, MD (Dalldorf Fellow, 1984)	
	D Stein, MD (Dalldorf Fellow, 1989)	
	A Waldorf, PhD	

A Sugar, MD

Degree	Postdoc	Mentees
	L Goldani, MD, PhD (Brazil)	
	A Angeles (one year in his lab)	
	ID Fellows (for 10 years)	

CENTERS FOR DISEASE CONTROL

Karling
↓

L Ajello, PhD (1947)

Degree	*Postdoc*	*Mentees*
	W Kaplan, DVM	W Kaplan
	L Kaufman, PhD	AF DiSalvo
	AF DiSalvo, MD	M McGinnis
	S Honbo, MD	M Rinaldi
	A Mantovani, DVM	
	T Matsumoto, MD	
	M Miyaji, MD	
	M McGinnis, PhD	
	L Morganti, DVM	
	L Polonelli, PhD	
	H Seeliger, MD	
	A Shoji, MD	

Benham
↓

L Georg, PhD (1950)

Degree	*Postdoc*	*Mentees*
	W Kaplan, DVM	W Kaplan
	C Jeffries	
	C Blank	

W Kaplan, DVM

Degree	*Postdoc*	*Mentees*
PhD Public Health students:	AF DiSalvo, MD	L Kaufman
(doing research projects)	L Kaufman, PhD	AF DiSalvo
B Reep		
E Williams		
B Phillips		
M Sudman		
W Turner		

CENTERS FOR DISEASE CONTROL (cont'd)

L Kaufman, PhD

Degree	Postdoc	Mentees
L Turner, PhD	F Odds, PhD	T Shinoda
K Jones, PhD	P Standard, PhD	L Mendoza
P Standard, PhD		
B Kleger, PhD		
J Brough, PhD		
C Sweet, PhD		
RB Lopez, PhD		
V Pideoe, PhD		

---

de Bary
↓
Farlow
↓
Thaxter
↓
Weston
↓
Conant, Wickerham, Lodder
↓   ↓

L Haley, PhD (1968–1986)

Degree	Postdoc	Mentees
L Stockman, MS	15 Postdoc students	M McGinnis
(Emory University)	(Histopath. and clinical)	
	M McGinnis, PhD	

Ainsworth
↓

Arvind A Padhye, PhD

Degree	Postdoc	Mentees
		M McGinnis, PhD

CENTERS FOR DISEASE CONTROL (cont'd)

Nickerson, Hasenclever
↓

E Reiss, PhD

Degree	*Postdoc (1976–1992)*	*Mentees*
RL Bradley, MS	R Hay, MD	CM Elie
GK Spearman, MS	PF Lehmann, PhD	
L Stockman, MS	L de Repentigny, MD	
LD Marr, MS	P Boiron, PhD	*Visiting scientists*
R Galindo, MS	TJ Lott, PhD	CD Jeffries, PhD
BJ Kemker, MS	H Diaz, MD	F Todaro-Luck, PhD
AE Jones, PhD	BA Lasker, PhD	OU Osoagbaka, PhD
F El-Zaatari, DrPH	GH Reyes, PhD	M Abu-samra, PhD
JR Knowles, PhD	RM Zancope-Olivera, PhD	Y Sugiura, PhD
KL Meckstroth, DrPH		
RM Zancope-Olivera, PhD		

Stevens
↓

C Morrison, PhD

Degree	*Postdoc*	*Sabbaticals*
	G Kimani, MD	S Fujita, MD/PhD
	S Zakroff, MD	JH Shin, MD/PhD
	M Lam, MD	L de Aguirre, DVM
	T Ward, DVM	
	D Kastin, MD	
	B Skaggs, PhD	
	K Copedge, DVM	

COLUMBIA UNIVERSITY

H Richards, PhD

Degree	*Postdoc*	*Mentees*
R Benham, MS, PhD (1919, 1931)		

Richards, Barnard, Dodge, Hopkins, Zinzer
↓

R Benham, PhD (1926–1955)
(College of Physicians & Surgeons)

Degree	*Postdoc/Visiting Scientists*	*Mentees*
LK Georg, PhD	B Kesten, MD	M Silva-Hutner
M Huppert, PhD	M Silva-Hutner, PhD	E Schnall
E Spitzer, MS, PhD	A Carrión, PhD	
	C Emmons, PhD	
	E Hazen, PhD 1944	
	S Rosenthal, PhD (1953)	
	M Hopper	
	L Ajello, PhD	
	E DeLamater, MD	

R Harper, PhD (Botany Chair)

Degree	*Postdoc*	*Mentees*
C Emmons, PhD (General mycology – 1931)		

JS Karling, PhD (Head, Mycology Department)

Degree	*Postdoc*	*Mentees*
L Ajello, PhD (1947)		L Ajello
J Sinski, PhD (at Purdue)		

155

COLUMBIA UNIVERSITY (cont'd)

de Bary
↓
Farlow
↓
Thaxter
↓
Weston
↓

M Silva-Hutner, PhD

Degree	*Postdoc*	*Mentees*
E Schnall, PhD	I Weitzman, PhD	M Belaval
	W Merz, PhD	I Weitzman
	J Torrez, Vet	W Merz
	R Travassos, MD	G Rebatta
	C Roitman, PhD	
	E Schnall, PhD	

Humphry
↓
Johnson
↓
Coker
↓
Couch
↓

L Olive, PhD (Botany Department)

Degree	*Postdoc*	*Mentees*
M Berliner, PhD		M Berliner
I Weitzman, PhD		
E Schnall, PhD		

A Carrión, PhD/G Hopkins, MD

Degree	*Research Assistant/Postdoc*	*Mentees*
M Belaval, MS	C Emmons, PhD (1929–1935)	M Silva-Hutner
		M Belaval

E Hazen, MA, PhD (1927)
Dept of Bacteriology,
Columbia University

CORNELL UNIVERSITY MEDICAL COLLEGE
Memorial Sloan-Kettering Cancer Center

D Armstrong, MD

Degree	Postdoc	Mentees
	B Wong, MD (1978–80)	

DUKE UNIVERSITY

DT Smith, MD (appointed Conant)

Degree	Postdoc	Mentees
H Hardin, PhD		

---

de Bary
↓
Farlow
↓
Thaxter
↓
Weston, Langeron, Sabouraud (Paris)
↓

N Conant, PhD (1935–1970)

Degree	Postdoc	Mentees	Visitors
L Haley, PhD	G Land, PhD	M Gordon	L Ajello
L Friedman, PhD	FT Wolf, PhD	H Gallis, MD	M Bartlett
(1951)	E Beneke, PhD	H Hardin, PhD	
C Halde, PhD	G Bulmer, PhD		
G Scherr, PhD	*H Larsh, PhD		
G Cozad, PhD (1953)			
R Vogel, PhD			
R Taylor, PhD			
R Hampson, PhD			
A Trejos, PhD			
I Christiansen, PhD			
*Information from his CV			

---

de Bary
↓
Farlow
↓
Thaxter
↓
Weston

FA Wolf, PhD (Botany-Mycology)

Degree	Postdoc	Mentees
M Gordon, PhD (1949)		
S McMillen, MA		

158

DUKE UNIVERSITY(cont'd)

de Bary
↓
Farlow
↓
Thaxter
↓
Weston
↓
Conant
↓
Friedman
↓

T Mitchell, PhD

Degree	*Postdoc*	*Mentees*
B Bolanos, PhD (1983)	S Lail, MD	
JM Small, PhD (1983)	B Ferguson, MD	
M Manning, PhD (1979)	W Meyer, PhD	
L Thurmond, PhD (1984)	R Li, PhD	

J Perfect, MD

Degree	*Postdoc/Fellows*	*Mentees*
	J Cohen	
	T Curiel	
	R Washburn	
	M Cairns	
	G Cox (current)	

D Durack, MD

Degree	*Postdoc*	*Mentees*
		J Perfect, MD

GEORGETOWN UNIVERSITY MEDICAL CENTER

Bessey
↓
Barnett, Blank
↓

R Calderone, PhD

Degree	*Postdoc*	*Mentees*
P Braun, PhD	M Fukayama, PhD	
	M Ollert, PhD	
	J Sturdevant, PhD	
	A Bailey, PhD	

HARVARD UNIVERSITY

de Bary
↓
Farlow[1]
↓
Thaxter
↓

W Weston,*[1,2] PhD 1915 (1921–1956)

Degree	Postdoc	Mentees
N Conant, PhD (1933)		
M Silva-Hutner, PhD		
FA Wolf, PhD		
R Emerson, PhD		
CM Wilson, PhD		
S Salvin, PhD		
A Howell, PhD		
R Page, PhD		
Total of 54 PhD students		

J Swartz, MD

Research Assistant

N Conant, PhD (1934–1935)

Humphry
↓
Johnson
↓
Coker
↓
Couch
↓
Olive
↓

C Campbell, BS[3] (Berliner[3] & Buckley[3])

Degree	Postdoc	Mentees
		A Restrepo
		M Berliner

*Contemporary of Couch, Dodge, Bessey, Wolf, and Farlow
[1]Sequeira L Annu Rev Phytopathol 1993; 31: 43–52
[2]Conant N Bull MMSA 1969; 7: 1–4
[3]"Tree" for Berliner; Campbell's mentees Harvard School for Public Health

IDAHO STATE UNIVERSITY

Levine
↓

GM Scalarone, PhD (1980)

Degree

MS	*MS*	*PhD*
R Abuodeh	D Towler	R Abuodeh
J Bono	A Wakamoto	J Bono
M Boyer	Y Wang	B Seawell
A Chamberlain	J Wheeler	
M Fisher	S Williams	
B Fryer	S Wisniewski	
M Hibbert	K Yearsley	
A Honigman	B Zaccardi	
D Houser		
S Johnson		
M Orr		
R Owens		
C Raman		
H Sato		
B Semple		
J Terry		

JOHNS HOPKINS MEDICAL INSTITUTE

Welch
↓

Benjamin Schenck, MD (late 1890's)

Caspar Gilchrist, MD (late 1890's)

Humphry
↓

D Johnson, PhD (1890's)

Degree	*Postdoc*	*Mentees*
W Coker, PhD (1897)		

Roger Baker, MD

---

de Bary
↓
Farlow
↓
Thaxter
↓
Weston
↓
Silva-Hutner
↓

W Merz, PhD (1970's)

Postdoc

D King, MD
T Walsh, MD

MEDICAL COLLEGE OF GEORGIA

J Denton, PhD, MD (1942)

Degree	*Postdoc*	*Mentees*
AF DiSalvo, MD		AF DiSalvo

MEDICAL COLLEGE OF OHIO

Kendrick
↓
Aldrich
↓
Alexopoulos
↓

G Cole, PhD

Degree	*Postdoc*	*Mentees*
	KR Seshan, PhD	
	JJ Yu	
	FE Williams, PhD	
	SL Smithson, PhD	
	PW Thomas, PhD	

METHODIST MEDICAL CENTER (Dallas, TX)

de Bary
↓
Farlow
↓
Thaxter
↓
Weston
↓
Conant
↓
Friedman
↓

G Land, PhD

Degree	*Postdoc*	*Mentees*
W Fleming		
A Sant		
H Prall		
T Davis		
S Thaxton		
R Roberts		
T Beadles		

165

MICHIGAN STATE UNIVERSITY

Shanor
↓

E Beneke, PhD (1948–1992)

Degree
AL Rogers, PhD
G Bulmer, PhD
R Sayer, PhD
K Steibel, MS, PhD
J Ueselenak, PhD

Postdoc
D Babel, PhD

Mentees
G Bulmer

Shanor
↓
Beneke
↓

AL Rogers, PhD

Degree
D Babel, PhD

Postdoc
D Hospinthal, PhD, MD
R Sandin, MS, MD
M Kennedy, PhD

Mentees

MONTANA STATE UNIVERSITY

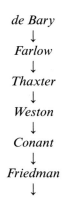

de Bary
↓
Farlow
↓
Thaxter
↓
Weston
↓
Conant
↓
Friedman
↓

JE Cutler, PhD (Tulane, 1972)

Degree	*Postdoc*
DL Brawner, PhD	DL Brawner, PhD
KC Hazen, PhD	T Kanbe, PhD
PM Glee, PhD	Y Han, PhD
R Li, PhD	T Caesar, PhD
Q Qian, PhD	P Glee, PhD
S Kobayashi, PhD	
R Morrison, MS	
A Poor, MS	
G Winterrowd, MS	
Five other MS students	

MYCO PHARMACEUTICALS, INC (Cambridge, MA)

Griffin
↓

W Timberlake, PhD

Degree	*Postdoc*	*Mentees*
W Orr	D Gwynne	
D Law	B Miller	
C Zimmermann*	E Mullaney	
J Hamer	T Adams	
P Mirabito	R Prade	
M Frizzell	Y Chang	
	A Andrianoupolos	
	R Fischer	

*At University of California, Davis

NATIONAL INSTITUTE OF ALLERGY and INFECTIOUS DISEASE (NIH) and NATIONAL CANCER INSTITUTE (NCI)

Harper, Carrión
↓

C Emmons, PhD (1936–1966)

Degree	*Postdoc*	*Mentees*
J Utz, MS (MD)	D Louria	J Bennett
	N Fedder	
	J Shadomy, PhD	
	J Bennett, MD	
	J Kwon-Chung, PhD	
	D Howard, PhD	
	H Hasenclever, PhD*	
	A Howell, PhD*	
	S Salvin, PhD*	
	L Pine, PhD*	

J Utz, MD, 1947

Degree	*Postdoc*	*Mentees*
	J Shadomy, PhD	J Shadomy

J Bennett, MD

Degree	*Postdoc*	*Mentees*
	JR Rhodes (Dalldorf Fellow, 1980)	W Dismukes
	J Rex, MD	J Rex
	V Kan, MD (Dalldorf Fellow, 1987)	

HF Hasenclever, Chief, MM section

Degree	*Postdoc*	*Mentees*
	E Reiss, PhD	

*C Campbell, personal communication, April 1, 1993

NATIONAL INSTITUTE OF ALLERGY and INFECTIOUS DISEASE (NIH) and NATIONAL CANCER INSTITUTE (NCI) (cont'd)

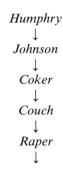

Humphry
↓
Johnson
↓
Coker
↓
Couch
↓
Raper
↓

J Kwon-Chung, PhD

Degree	*Postdoc*	*Mentees*
B Wickes, PhD	S Kim, PhD	B Wickes
(at Catholic University)	K Min, PhD	G Bulmer
	J Rhodes, PhD (Dalldorf Fellow, 1980)	
	I Polacheck, PhD	
	S Kahno, PhD	
	W Whelan, PhD	
	PT Magee, PhD	

**G Bodey, MD, PA Pizzo, MD,
TJ Walsh, MD at NCI**

Mentees		
W Venanzi, MD	C McEntee, PhD	R Schaufele, MS
S Aoki, MD	P Francis, MD	M Hom-Eng, MD
F Mechinaud, MD	K Ühlig, MD	A Gehrt, MD
JW Lee, MD	M Allende, MD	M Lasso, MD
E Roilides, MD	C Gonzalez, MD	N Ali, MD
R Pelletier, MD	T Sein, MD	C Lyman, PhD
J Berenguer, MD	C Blake, MD	A Holmes, MD
J Lecciones, MD	K Garrett, MS	NV Thomas, DDS
E Navarro, MD	P Kelly, MD	D Arenberg, MD
S Ravankar, MD	K McNight, DDS	A Hsia, MD
S Lee, MD	M Ashe, PhD	R Torrez, MD
A Francesconi, MS		

NEW YORK STATE HEALTH DEPARTMENT

E Hazen, PhD
Fungus lab from NY to Albany (1944)

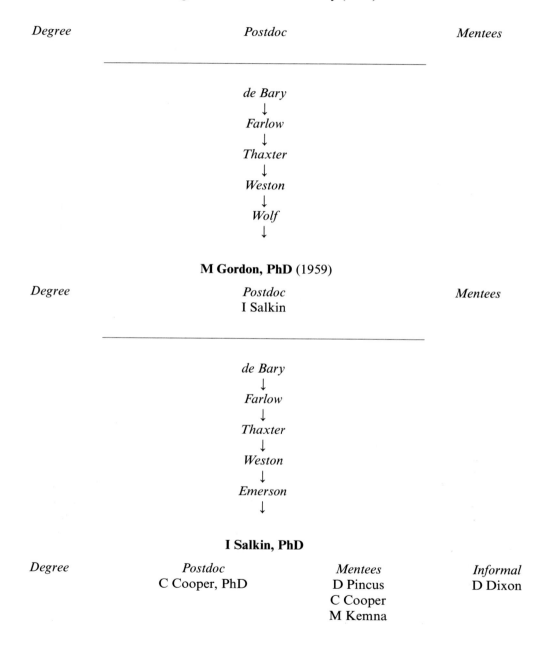

Degree	*Postdoc*	*Mentees*

de Bary
↓
Farlow
↓
Thaxter
↓
Weston
↓
Wolf
↓

M Gordon, PhD (1959)

Degree	*Postdoc* I Salkin	*Mentees*

de Bary
↓
Farlow
↓
Thaxter
↓
Weston
↓
Emerson
↓

I Salkin, PhD

Degree	*Postdoc* C Cooper, PhD	*Mentees* D Pincus C Cooper M Kemna	*Informal* D Dixon

NEW YORK STATE HEALTH DEPARTMENT (cont'd)

Plunkett
↓
J Shadomy
↓

D Dixon, PhD

Degree	Postdoc	Mentees	Informal
	C Cooper, PhD		
	T Walsh, MD		

RUTGERS UNIVERSITY(New Brunswick, NJ)

WJ Nickerson, PhD

Degree	*Postdoc*	*Mentees*
VK Jansons, PhD(1968)		E Reiss
S Bartnicki-Garcia, PhD (1970)		
E Reiss, PhD (1972)		

SAINT LOUIS UNIVERSITY AT ST LOUIS, MO

de Bary
↓
Farlow
↓
Thaxter
↓
Weston
↓
Conant
↓
Friedman
↓
Kobayashi
↓

E Keath, PhD

Degree	*Postdoc*	*Mentees*
A Ignatov	C Weaver	
F Abidi	H Roh	

SANTA CLARA VALLEY MEDICAL CENTER

D Stevens, MD and Research Associates: Brummer* and Clemons*

Degree	Postdoc	Mentees
J Byron, MS	J Galgiani, MD	A Sugar
S Duvvuru, MS	A Sugar, MD	J Galgiani
E Castañeda, PhD	K Clemons, PhD	K Clemons
K Kamei, PhD (MD)	D Denning, MBBS	D Denning
	C Morrison, PhD	C Morrison
Students	D Dewsnup, DO	
S Park	R Tucker, MD	
J Labson	E Arathoon, MD	
P Chahal	J Hostetler, MD	
D Robertson	C Brass, MD	
L Blaine	P Williams, MD	
S Hurley	R Stiller, MD	
L Hanson	E Lefler, PhD	
A Perlman	A Ganer, MD	
T Tran	J Shainhouse, MD	
Y Minn	S Deresinski, MD	
S Crosby	A Zolopa, MD	
C Cox	R Harvey, MD	
L Catena		
D Hermoso		
T Drasin		
N Randhawa		

Guthrie, Cox
↓

*E Brummer, PhD (1978)

Student	Postdoc	Mentees
E Castañeda, PhD	K Kurita, PhD	
K Kamei, PhD (MD)	K Kamei, MD, PhD	
	J McEwen, MD, PhD	
	P Morozumi, MD	
	F Nassar, MD	

*K Clemons, PhD

Student	Postdoc	Mentees
B Aristizabal	M McCullough, PhD; DDS	
P Park		

TEMPLE UNIVERSITY

R Sabouraud
↓

F Blank, PhD/S Grappel, PhD

Degree	*Postdoc*	*Mentees*
C Abramson, PhD (1965)	R Hopfer, PhD (1972–74)	R Hopfer
(+ H Friedman)	R Calderone, PhD (1973–74)	C Abramson

H Buckley, PhD (1977–2001)

Degree	*Postdoc*
MS degree: Seven students	Six fellows
PhD candidates: 13 students	

TULANE UNIVERSITY

de Bary
↓
Farlow
↓
Thaxter
↓
Weston
↓
Conant
↓

L Friedman, PhD (1955–1980)

Degree	*Postdoc*	*Mentees*
G Land, PhD	J Sinski, PhD (1958–60)	A Restrepo
T Mitchell, PhD (1971)	J Held, PhD	J Domer
G Kobayashi, PhD	P Smith, PhD	G Land
J Cutler, PhD	R Tesh, MD	J Sinski
W Cooper, PhD		J Murphy
M Dykstra, PhD		D Pappagianis
K Erke, PhD		J Cutler
S Ford, PhD (1965)		
D Greer, PhD (1964)		
S Kraeger, PhD		
A Bonk, MS		
M Ivens, MS		
L Linares, MS		
M Martin, MS		

M Shaffer

Degree	*Postdoc*	*Mentees*
JD Schneidau, PhD (1957)		
J Domer, PhD (1966)		

Shaffer
↓

J Schneidau Jr, PhD

Degree	*Postdoc*	*Mentees*
A Restrepo, MS, PhD		

176

TULANE UNIVERSITY (cont'd)

Shaffer
↓

J Domer, PhD

Degree	Postdoc	Dissertation Committee
D Giger, PhD	S Moser	G Land
T Davis, PhD	F Lyon	M Dykstra
E Carrow, PhD	R Hector	R Sakulramrung
G Andersen, PhD	R Garner	P Lammie
V Kuruganti, MS	A Childress	A Henderberg
	L Al-Hussaini	
	Y Wang	

Note: Also at Tulane, M Littman, 1945–1948, who trained Schneidau. Information on Held, Smith, Tesh, and PhD, MS students below G Kobayashi was gathered from MF Shaffer's "History of Medical Mycology at Tulane Medical School", 1955. Information was not available for Shaffer's major professor.

UNIVERSITY OF ALABAMA

W Dismukes, MD

Degree	*Postdoc*	*Mentees*
	M Saag, MD	T Kerkering
	P Pappas, MD	M Saag
	A Stamm, MD	
	M Pierce, MD	
	S Alsip, MD	
	G Karam, MD	

UNIVERSITY OF ARIZONA

Karling
↓

JT Sinski, PhD (1966)

Degree	*Postdoc*	*Mentees*
G Reed, MS		
J Swanson, MS		
P Flynt, MS, PhD		
V Wallis, MS		
J Pokrifchak, MS		
T Moore, MS		
D Van Avermaete, MS		
S Kutz, MS		

VA MEDICAL CENTER

J Galgiani, MD

UNIVERSITY OF CALIFORNIA AT BERKELEY

CE Smith, MD (1949–1967)

Degree	*Postdoc*	*Mentees*
D Pappagianis, PhD, MD	L Friedman, PhD (Research mycologist) 1951–1955	

---

de Bary
↓
Farlow
↓
Thaxter
↓
Weston
↓

R Emerson, PhD (Botany Dept)

Degree	*Postdoc*	*Mentees*
I Salkin, PhD (1969) A Held, PhD D Griffin, PhD M Tansey, PhD M Fuller, PhD H Whisler, PhD		

H Phaff, PhD

Degree

H Levine, PhD

Degree
G Scalarone, PhD
M Valesco, DPH

UNIVERSITY OF CALIFORNIA AT DAVIS

CE Smith
↓

D Pappagianis, MD, PhD

Degree	*Postdoc*	*Mentees*
M Collins, PhD		M Rinaldi
R Hector, PhD		
B Zimmer, PhD		
S Johnson, PhD		
E Castañeda, PhD		
A Ibrahim		

P Hoeprich, MD

Degree	*Postdoc*	*Mentees*
M Rinaldi, MS, PhD (1980)	M Rinaldi, PhD	
M Saubolle, PhD		

UNIVERSITY OF CALIFORNIA AT LOS ANGELES

O Plunkett, PhD (Botany Dept, 1929)

Degree	*Postdoc (Visiting)*	*Mentees*
C Halde, MA (1947)	JW Wilson, MD	D Howard
BG Studt, PhD		C Halde
R Cutter Burke, PhD		J Shadomy
J Taggart, PhD		S Shadomy
F Swatek, PhD		
HJ Shadomy, PhD		
R Wishard, PhD		
G Orr, PhD		
H Walsh, PhD		
R Peterburg, PhD		
S Pore, PhD		
F Ramsdile, PhD		
P Martin, PhD		
J Briggs, MA		
L Miller, MA		
R Yamamoto		

J Kessel

Degree	*Postdoc*	*Mentees*
EE Evans, PhD		

G Jann (Bacteriology Dept)

Degree	*Postdoc*	*Mentees*
D Howard, PhD (1954)		

D Howard, PhD

Degree (1964–1993)	*Postdoc*	*Mentees*
N Dabrowa, PhD (1968)	I Barash, PhD	M Clancy, MS
J Garcia, PhD (1968)	S Gupta, PhD	
J Yen, PhD (1969)	J Yen, PhD	
B Gilbert, PhD (1970)	N Dabrowa, PhD	
S Tang, PhD (1973)	S Chui, PhD	
J Rhodes, PhD (1980)	D Oblack, PhD	
B Wu-Hsieh, PhD (1983)	J Rhodes, PhD	
	B Wu-Hsieh, PhD	
M Zeuthen, MS, PhD (1989)	L Nakamura	
L Villarete, PhD (1993)	(Summer medical students	
Burnett, MS (1968)	1958–1973)	
Scharlin, MS (1969)		
S Munnerlyn, MS (1973)		
G Gibbons, MA (1983)		

UCLA HARBOR Medical Center

J Edwards, MD

Degree	Postdoc	Mentees
	S Filler, MD	J Rex, MD
	(Dalldorf Fellow, 1988)	

UNIVERSITY OF CALIFORNIA AT SAN FRANCISCO

N. Conant
↓

C Halde, PhD

PhD Committees	Postdoc	Mentees
M Valesco, DPH	A Poperoff, MD	M McGinnis
A Waldorf, PhD		
E Castañeda, PhD		

J Edman, PhD

PhD Degree	Postdoc	Mentees
	B Wickes, PhD	
	A Schlageter, PhD	

H Larsh
↓

R Aly, PhD

PhD Degree	Postdoc	Mentees
		J Fa-Geman, MD

N Agabian

Degree	Postdoc	Mentees
S Miyasaki, DDS, PhD	A Kuo, PhD	G Newport, PhD
	T White, PhD	

T White, PhD

UNIVERSITY OF CHICAGO

J Rippon, PhD

Degree	Postdoc	Mentees
		D Babel
		I Salkin

UNIVERSITY OF CINCINNATI COLLEGE OF MEDICINE

J Schwarz, MD (1948) & G Baum, M D

Degree	Postdoc	Mentees
	C Kauffman, MD	J Rhodes

W Bullock, MD

Degree	Postdoc	Mentees
	G Deepe, MD	
	(Dalldorf Fellow, 1981)	

G Deepe, MD

Degree	Postdoc	Mentees
AM Gomez, MD, PhD	F Gomez, MD	
	R Allendoerfer, MD	

J Rhodes, PhD

Degree	Postdoc	Mentees
AM Gomez, MD, PhD	DR Bode, MD	
	S Kralovic, MD	

P Walzer, MD

Degree	Postdoc	Mentees
M Linke, PhD	M Cushion, PhD	M Cushion
	S Theus, PhD	M Linke
	G Smulion, MD	S Theus
		S Smulion

B Wong, MD (1984)

Degree	Postdoc	Mentees
	J Murray, MD	
	V Chaturvedi, PhD	
	J Wang, MD	

UNIVERSITY OF ILLINOIS

de Bary
↓
Farlow
↓
Thaxter
↓
Weston
↓
Conant
↓

G Scherr, PhD

Degree	*Postdoc*	*Mentees*
J Rippon, MS, PhD		

L Shanor, PhD

Degree	*Postdoc*	*Mentees*
E Beneke, PhD		

UNIVERSITY OF IOWA

G Martin (classic mycologist)

Degree	Postdoc	Mentees
C Emmons, MS (1927)		
WA Taber, PhD		

Richards
↓
Benham
↓
Huppert
↓

J Cazin, PhD

Degree	Postdoc	Mentees
T Kozel, PhD		
D Lupan, PhD		
D Kitz		
K Anderson		
W Burt		

D Soll, PhD

Degree	Postdoc	Mentees
L Mitchell, MS	G Bedell	
B Kraft, PhD	W Whelan	
J Hellstein, MS	R Finney	
B Morrow, PhD	B Varnum	
C Kleinegger, MS	B Slutsky	
K Vargas, DDS	TVG Rao	
Current Graduate Students	J Anderson	
L Enger	M Bergen	
H Ramsey	J Schnid	
A Sturdevant	E Voss	
C Kvaal	T Srikantha	
M Herman	S Lockhart	
C Pujol	K Schroppe	
S Joly		

M Pfaller, MD

Degree	Postdoc	Mentees
	T Perl, MD	
	B Doebbeling, MD	
	D Reagan, MD	
	A Pignatari, MD	
	M Branchini, MD	
	M Sabria, MD	

UNIVERSITY OF KENTUCKY

Larsh
↓

N Goodman, PhD, ML Furcolow, MD

Degree*	*Postdoc*	*Mentees*
H Hempel, MS (1974)	G Roberts, PhD (1972)	J Snyder
D Schwartz, MS (1974)	J Susilo, MD (1973)	E Koneman
R Barnett, MS (1978)	J Darner, PhD (1973)	
E Hempel, MS (1978)	R Scarry, PhD (1974)	
B Humbard, MS (1979)	R Otero, PhD (1975)	
M Parek, MS (1981)	L Phillips, PhD (1975)	
A Johnson, MS (1981)	A Hairi, PhD (1979–81)	
B Zimmer, MS (1982)	A Brown, PhD (1984)	
S Kelly, MS(1982)	R Victoric, MD (1991)	
M Brooks, MS (1985)		
G Hawkins, MS (1985)		
D Riyonto, MS (1987–88)		

*Goodman's students

UNIVERSITY OF MINNESOTA COLLEGE OF BIOLOGICAL SCIENCES
(St Paul, Minnesota)

PT Magee, PhD

Degree	*Sabbatical/Postdoc*	*Informal Training*
M Osley	J Perfect, MD	J Rhodes, PhD
L Hereford	B Wong, PhD	L Chaffin, PhD
L Smith	J Suzuki, PhD	
S Suhny	B Whelan, PhD	
W Chu		

UNIVERSITY OF NEVADA

Richards
↓
Benham
↓
Huppert
↓
Cazin
↓

T Kozel, PhD

Degree	*Postdoc*	*Mentees*
T McGaw		
F Swanson		
A Schlageter, PhD		

UNIVERSITY OF NORTH CAROLINA AT CHAPEL HILL

Humphry
↓
Johnson
↓

W Coker, PhD (1901)

Degree	*Postdoc*	*Mentees*
J Couch, PhD (1924)		
HR Totten		
L Olive, PhD		
V Matthews		
L Shanor		

Note: W Coker, a Botanist/Mycologist, received his PhD from Johns Hopkins University in 1897. His mentor there was the Botanist, D Johnson (PhD 1897, Johns Hopkins), who was a student of J Humphry (Botanist) at Johns Hopkins (P Szaniszlo, Personal Communication, 1993).

Humphry
↓
Johnson
↓
Coker
↓

JN Couch, PhD (1927)

Degree	*Postdoc*	*Mentees*
K Raper, MA (1936)		J Raper, MA
A Lang, PhD (1936)		25 other MA students
G Christenberry, PhD		
J Doubles, PhD		
A Wiffin, PhD		
A Ziegler, PhD (1941)		
E Goldie-Smith, PhD (1953)		
M Huneycutt, PhD (1955)		
W Koch, PhD (1956)		
C Miller, PhD (1957)		
J Mullins, PhD (1959)		
C Umphlett, PhD (1961)		
M Slifkin, PhD (1967)		
P Szaniszlo, PhD (1967)		
W Sherwood, PhD (1967)		
C Bland, PhD (1968)		

Note: Information for Dr Couch's students (except K Raper and P Szaniszlo) was obtained from: Szaniszlo, P (1993). John Nathaniel Couch 1896–1986. Biographical Memoirs, 62, Washington, DC: The National Academy Press.

UNIVERSITY OF NORTH CAROLINA AT CHAPEL HILL (cont'd)

de Bary
↓
Farlow
↓
Burt
↓
Gilman
↓
Tiffany
↓

M R McGinnis, PhD

Degree	*Postdoc*	*Mentees*
		M Rinaldi
		I Salkin
		L Pasarell

Committee Member
B Katz, PhD
W Schell, MS
K Hastings, DPH

Sabouraud
↓
Blank
↓

R Hopfer, PhD (1989)

UNIVERSITY OF OKLAHOMA (Norman & Oklahoma City)

H Larsh, PhD
(Norman, 1941, 1945)

Degree	*Postdoc*	*Mentees*
J Greer, PhD (1956)	N Goodman, PhD	A Guruswamy
T Mahvi (1961)	RA Cox, PhD	J Murphy
R Ritter, MS, PhD (1961)	R Pore	N Goodman
N Goodman, PhD (1965)	A Morehart	C Lyman
R Sprouse (1966–1967)	F Cantrell	
P Bartels, PhD (1968)	W Sorenson	
S Fadula, PhD (1969)		
R Meyer, PhD		
R Aly, PhD (1969)		
G Roberts, PhD (1972)		
B Frye, PhD (1972)		
R Unger, PhD (1972)		
R Karaoui, PhD (1975)		
K Young, MS, PhD (1978-1981)		
G Hollick, PhD (1979)		
D Schwartz, PhD (1979)		
R Longley, PhD (1981)		
S Landon-Arnold, PhD (1982)		
R Morrison, PhD (1982)		
G Shearer, PhD (1983)		
N Hall, MS		
C Lyman, MS		

M Furcolow, MD/Larsh, PhD

Degree	*Postdoc*	*Mentees*
G Cozad, MS (1954)		

UNIVERSITY OF OKLAHOMA (Norman & Oklahoma City) (cont'd)

de Bary
↓
Farlow
↓
Thaxter
↓
Weston
↓
Conant
↓

G Cozad PhD (Norman)

Degree	*Postdoc*	*Mentees*
J Murphy, PhD (1969)		
D Adamson		
D Spencer		
S Siddiquis		
J Rhodes, MS (1973)		
G Scalarone, MS		

Shanor
↓
Beneke
↓

G Bulmer, PhD (Oklahoma City)

Degree	*Postdoc*	*Mentees*
R Fromtling, PhD (1979)		R Fromtling
A Ruiz, PhD (1981)		
B Robinson, PhD (1981)		
R Tacker, PhD, MD (1975)		
F Farhi, PhD (Teheran) (1977)		
M Morgan (1982)		

H Muchmore, PhD (Oklahoma City)

Degree	*Postdoc*	*Mentees*
	R Fromtling, PhD	

UNIVERSITY OF OKLAHOMA (Norman & Oklahoma City) (cont'd)

de Bary
↓
Farlow
↓
Thaxter
↓
Weston
↓
Conant
↓
Cozad
↓

J A Murphy, PhD (Oklahoma City)

Degree	*Postdoc*	*Mentees*
P Fidel, MS, PhD	K Howard, PhD	P Fidel
T Lim, PhD	K Buchanan, PhD	
N Nabavi, PhD	B Jimenez-Finkel, MS, PhD	
P Fung		
B Jimenez-Finkel, MS, PhD		
K Buchanan, MS		
ZM Dong, PhD (1995)		
+17 other MS students		

UNIVERSITY OF TEXAS

D Drutz, MD
(1970s at San Antonio)

Degree	Postdoc	Mentees
	S Klotz , MD	C Frey, Msc (1980–90)

J Graybill, MD
(San Antonio)

Degree	Postdoc	Mentees	Worked in Lab
			P Craven, MD
			B Restrepo
			M Patino
			A Gomez
			M Robledo
			F Moreno
			B Atkinson

Hoeprich
↓

M Rinaldi, PhD
(San Antonio)

Degree	Postdoc	Mentees	Informal
		R Larsen, MD	>30 individuals have
		A Fothergill, BS	been in his laboratory

R Cox, PhD
(San Antonio Chest Hospital)

Degree	Postdoc	Mentees
	E Brummer, PhD	

193

UNIVERSITY OF TEXAS (cont'd)

de Bary
↓
Farlow
↓
Thaxter
↓
Weston
↓
Conant
↓
Friedman
↓

B Cooper, PhD (Dallas)

Degree	*Postdoc*	*Mentees*
		G Land

Humphry
↓
Johnson
↓
Coker
↓
Couch
↓

P Szaniszlo, PhD (Austin)

Degree	*Postdoc/Visiting Scholars*	*Mentees*
R Roberts, PhD (1979)	S Grove, PhD	C Cooper, PhD
P Geis, PhD (1981)	P Powell, PhD	S Chua, BS
C Jacobs, PhD (1983)	J Harris, PhD	B Krom
C Cooper, PhD (1989)	D Crowley, PhD	
M Nelson, PhD (1991)	A Thalaseedharan, PhD	
M Momany, PhD (1992)	S Karuppayil, PhD	
Y Wang, PhD (1993)	M Peng, PhD	
L Mendoza, PhD (1995)	Z Yin, MD, PhD	
L Zheng (1997)		
Z Wang (1998)		
X Ye (1998)		
+ 14 MS students		

UNIVERSITY OF TEXAS (cont'd)

Kendrick
↓
Aldrich
↓
Alexopoulos
↓

G Cole, PhD (Austin)

Degree	*Postdoc*	*Mentees*
D Kruse, PhD	J Pishko, PhD	
L Yuan, PhD	E Wyckoff, PhD	
L Pope, PhD	PW Thomas, PhD	
S Zhu, PhD	SL Smithson, PhD	
S Pan, PhD	KR Seshan, PhD	
JJ Yu, PhD		
K Tanaka, MS		
W Cawley, MS		
C Williams, MS		

---

de Bary
↓
Farlow
↓
Burt
↓
Gilman
↓
Tiffany
↓

MR McGinnis, PhD (Galveston)

Degree	*Postdoc*	*Mentees*
A Espinel-Ingroff, PhD (at MCV, VCU)		C Cooper, PhD
		L Pasarell, MPH

MD ANDERSON CANCER CENTER (Houston)

G Bodey, MD

Degree	Postdoc	Mentees
		E Anaissie, MD

E Anaissie, MD

Degree	Postdoc	Mentees
	D Kontoyiannis, MD	
	D Kohen, MD	
	C Legrand, MD	
	A Gokaslan, MD	

RC Hopfer, PhD (1974–87)

Degree	Postdoc	Mentees
		G Araj, PhD

UNIVERSITY OF VIRGINIA

de Bary
↓
Farlow
↓
Thaxter
↓
Weston
↓
Conant
↓
Friedman
↓
Cutler
↓

K Hazen, PhD

Degree	Postdoc	Mentees
	PM Glee, PhD	
	TJ Silva, MD	
	J Masuoka, PhD	

UNIVERSITY OF WASHINGTON (Seattle) and
SEATTLE BIOMEDICAL RESEARCH INSTITUTE

T White, PhD

Degree	*Postdoc*	*Mentees*
	JoBeth Harry, PhD	K Marr, MD

FRED HUTCHINSON CANCER RESEARCH CENTER (Seattle)

R Bowden, MD

Degree	*Postdoc*	*Mentees*
		K Marr, MD
		Jo van Burik, MD

UNIVERSITY OF WISCONSIN

Humphry
↓
Johnson
↓
Coker
↓
Couch
↓

K Raper, PhD

Degree	*Postdoc*	*Mentees*
J Kwon-Chung, MS, PhD		M Berliner

J Davis/J Jones, MD
(Hospital & Clinics)

Degree	*Postdoc*	*Mentees*
	B Klein, MD	
	(Dalldorf Fellow, 1986)	

EM Gilbert

Degree	*Postdoc*	*Mentees*
FT Wolf, PhD		

VA MEDICAL CENTER – UNIVERSITY OF MICHIGAN
(Ann Arbor, Michigan)

C Kauffman, MD

Degree	Postdoc/Fellows	Mentees
	J Sangeorzan, MD	
	S Bradley, MD	
	P Jones, MD	
	R Tiballi, MD	

VIRGINIA COMMONWEALTH UNIVERSITY

Plunkett
↓

S and J Shadomy, PhD (1960s)

Degree	*Postdoc*	*Mentees*
D Dixon, PhD	J Fisher, MD	R Gebhart, MD
B Davis, PhD	L Austin, PhD	C Utz, MS
S Elmorshidy, MS	R Duma, MD	J Fisher, MD
P Goldson, MS	R Fromtling, PhD	
C Utz, MS	R Gebhart, MD	
L Nicholas, MS	T Kerkering, MD	*Informal Training*
	(Dalldorf Fellow, 1978)	A Espinel-Ingroff, PhD
	N Rotowa, MD	

Harper
↓
Emmons
↓

J Utz, MD, MS (1960s)

Degree	*Postdoc*	*Mentees*
	J Shadomy, PhD	J Shadomy
	R Duma, MD	T Kerkering
		R Duma

R Duma, MD, PhD

Degree	*Postdoc*	*Mentees*
		T Kerkering

VIRGINIA COMMONWEALTH UNIVERSITY (cont'd)

A Espinel-Ingroff, MS, PhD

Degree	Postdoc	Mentees	Informal Training
			S Elmorshidy
			D Dixon
			J Fisher, MD
			T Kerkering, MD
			R Gebhart, MD
			M Tenenbaum, MD
			M Turik, MD
			V Bruzzese, MD
			M Workman, MD
			T Flynn, MD

Metzenberg
↓

E Jacobson, MD, PhD (VA Hospital)

Degree	Postdoc	Mentees
C White, PhD	S Vartivarian, MD	
R Dent, MS	H Emery, PhD	
C Still, MS	K Nyhus, PhD	
	R Ikeda, PhD (Sabbatical)	

T Kerkering, MD

Degree	Postdoc	Mentees
	M Turik, MD	

WASHINGTON UNIVERSITY SCHOOL OF MEDICINE

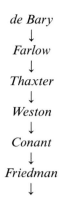

de Bary
↓
Farlow
↓
Thaxter
↓
Weston
↓
Conant
↓
Friedman
↓

G Medoff, MD/Kobayashi, PhD

Degree	*Postdoc*	*Mentees*	*Informal*
	B Goldman	M Pfaller	E Jacobson, MD
	B Lasker, PhD		(1975–77) (Fellow)
	P Steele		M Pfaller, MD
	E Keath		(1980–82)
	E Spitzer		
	J Arroyo		
	B Vincent		
	J Wolf		
	B Maresca, PhD		
	M Sacco		
	D Kitz		
	B Body, PhD		

G Medoff, MD

Degree	*Postdoc*	*Mentees*
	M Sokol-Anderson, MD	
	(Dalldorf Fellow, 1985)	

WAYNE STATE UNIVERSITY (Detroit, MI)

Reiss
↓

C Jeffries, PhD

Degree*	*Postdoc*	*Mentees*
P Bajwa, PhD		
P Whitcomb, PhD		

* Medical Mycology Thesis (PhD) but they are not working in medical mycology

---

J Sobel, MD

Degree	*Postdoc*	*Mentees*

EK Manavathu, PhD

Degree	*Postdoc*	*Mentees*

JA Vazquez, MD

Degree	*Postdoc*	*Mentees*

WEST VIRGINIA UNIVERSITY

Bessey
↓

H Barnett

Degree	*Postdoc*	*Mentees*
W Merz, MS, PhD (1968)	R Calderone	
R Calderone, PhD		
C Kurtzman, PhD		

YALE UNIVERSITY SCHOOL OF MEDICINE
VA Connecticut Health Care System, West Haven, CT

B Wong, MD

Postdoc
HT Truong, MD (1996)
Y Mao, PhD (1996)

V Chaturvedi, PhD (Assoc. Research Scientist)

SELF-TAUGHT (Not "formally" taught)

Davise Larone, PhD
Ted White, PhD

Appendix C: Medical mycology books (chronologic order)

Jacobson HP. Fungous Diseases: A Clinico-mycological Text. Springfield, Il: Charles C. Thomas, 1932.
"The first book in the U.S.A. that dealt solely with medically important fungi. It is strictly descriptive in a clinical sense and does not discuss etiological agents" (G Kobayashi, personal communication, July 1, 1993).

Dodge CW. Medical Mycology: Fungous Diseases of Men and Other Mammals. St. Louis: C.V. Mosby Co., 1935.
This is a very significant book. "Although predated by HP Jacobson's book, the book by Dodge is the first complete treatise describing all of the fungi that have been medically implicated in disease up to then" (G Kobayashi, personal communication, July 1, 1993; M McGinnis, questionnaire).

Lewis GM, Hopper ME. An Introduction to Medical Mycology. Chicago: The Yearbook Publishers, Inc., 1939.
This book and other early ones "put medical mycology on a sound scientific basis and organized the information rationally" (D Pappagianis, questionnaire).

Conant NF, Martin DS, Smith DT, Baker RD, Callaway JL. Manual of Clinical Mycology. Philadelphia: W.B. Saunders Co., 1944 (2nd edn., 1954; 3rd edn, 1971).
This book "cleared away the seeming disheartening complexity of the field and the miasma of misinformation" (Ajello, personal communication, December 23, 1993).

Skinner CE, Emmons CW and Tsuchiya HM. Molds, Yeasts, and Actinomycetes. London: John Wiley & Sons, Inc., 1947.
"This book is very important. While listed as the second edition of Henrici's first book (published in 1930), Henrici did not write this edition. It was written in memory of this great teacher of mycology by the three listed authors". It is the basis of the book by Emmons CW, Binford CH, Utz JP (1963) entitled, Medical Mycology. Philadelphia: Lea & Febiger (G Kobayashi, personal communication, July 1, 1993).

Wolf FA, Wolf FT. The Fungi in Two Volumes. New York: John Wiley & Sons, Inc., 1947.
Volume II contains the medical mycology section.

Wilson JW. Clinical and Immunological Aspects of Fungous Diseases. Springfield, Ill: Charles C. Thomas, 1957.
This is a well known book since it contains the first description of the "primary chancriform syndrome" as opposed to cutaneous lesions resulting from disseminated systemic disease (F Swatek, personal communication, January 10, 1994).

Emmons CW, Binford CH, Utz JP. Medical Mycology. Philadelphia: Lea & Febiger, 1963.
Emmons edited three editions and in 1977, J. Kwon-Chung became an author (3rd edn.). Kwon-Chung and J Bennett published a fourth edition in 1992.

Ajello L, Georg LK, Kaplan W, Kaufman L. Laboratory Manual for Medical Mycology. Atlanta: Public Health Service Publication No. 994, CDC, 1963.

Wilson JW, Plunkett OA. The Fungous Diseases of Man. Berkeley & Los Angeles: University of California Press, 1965.
This book was a very good teaching tool as it included a useful series of drawings and photographs of medically important fungi.

Haley LD. Diagnostic Medical Mycology. New York: Appleton-Century-Crofts, 1965.
Haley first published this manual at Yale University. Two more editions were published (1973, 1978) while she was at CDC. This manual made "possible to do in the laboratory what Conant and coauthors (1944) have stated was important to do in order to diagnose fungal diseases and to identify the fungi causing them" (LD Haley, personal communication, December 31, 1993).

Beneke ES. Monograph in Human Mycoses. Kalamazoo, MI: The Upjohn Company, 1968.
Seven editions have been published of this manual. It was an excellent teaching tool for medical students for many years at the VCU/Medical College of Virginia.

Rebell G, Taplin D. Dermatophytes. Their Recognition and Identification. Coral Gables, FL: University of Miami Press, 1970.
This manual provided an aid in the identification of the etiologic agents of skin and hair infections. It was published first in 1964 in photo-offset with three color plates by the Dermatology Foundation of Miami. Since the first edition, the number of recognized members of the genera Epidermophyton, Microsporum, and Trichophyton had increased from 28 to over 40. Therefore, the 1970 edition was designed to include these additional species and had a new color plate showing the new ascigerous species of *Microsporum* and *Trichophyton*.

Rippon JW. Medical Mycology. Philadelphia: W.B. Saunders Co., 1974 (2nd edn., 1982; 3rd edn., 1988).
The three editions provide a compendium of information used in the training of physicians and laboratory personnel.

McGinnis MR. Laboratory handbook of medical mycology. New York: Academic Press, 1980.
"This handbook provided the first usable bench manual, which included a listing of all previously reported fungal pathogens" (I Salkin, personal communication, April 25, 1994). The manual "contains established clinical nomenclature and contemporary mycological terminology" (M McGinnis, personal communication, September 17, 1993).

Chandler FW, Kaplan W, Ajello L. Color Atlas and Text of the Histopathology of Mycotic Diseases. Chicago: Year Book Medical Publishers, Inc., 1980.
This book included a vast selection of photomicrographs that represent fungal lesions with corresponding laboratory diagnostic captions to facilitate the rapid review or differential diagnosis of a fungus in tissue.

204

Howard DH, ed. Fungi Pathogenic for Human and Animals (in three parts). New York: Marcel Dekker, Inc., 1983.

Szaniszlo PJ, ed. Fungal Dimorphism: With Emphasis on Fungi Pathogenic for Humans. New York: Plenum Publishing Corporation, 1985.
"The first complete volume of information devoted exclusively to the subject of dimorphism" (PJ Szaniszlo, personal communication, June 6, 1994). It provided complete reviews of dimorphism of a large variety of fungi that are mostly pathogenic for humans and other animals, e.g. *B. dermatitidis*, *H. capsulatum*, *C. albicans*, and *C. immitis*.

Reiss E. Molecular Immunology of Mycotic and Actinomycotic Infections. New York: Elsevier Biomedical Pubications, 1986.

Kwon Chung KJ, Bennett JE. Medical Mycology. Philadelphia: Lea & Febiger, 1992.

This book focused on fundamentals of interest to most medical mycologists.

Larone DH. Medically Important Fungi. A Guide to Identification, 2nd edn. Washington, D.C.: American Society for Microbiology, 1993.
This unique and extremely popular book answers the demand of laboratory personnel in need of reliable and simple guidelines for identifying fungi that are clinically encountered. The material is arranged so that the laboratorian can systematically reach a possible identification knowing only the macro- and microscopic morphology of an isolated organism The third edition was published in 1995 by the American Society for Microbiology.

Murphy J, Friedman H, Bendinelli M, eds. Fungal Infections and Immune Response. New York: Plenum Press, 1993.
This book presents recent advances in the understanding of fungal pathogens.

Appendix D: Historical publications by other authors

Biographic studies

Silva and Hazen (1957) and Kwon-Chung and Campbell (1986) have published biographic summaries of Rhoda Benham and Chester Emmons. Benham was a professor and mentor of the former authors and Emmons was a mentor for the latter authors. Benham's scientific publications, as well as her influence as a teacher, are considered her major contributions to medical mycology. Benham's contributions as a pioneer researcher and first leader in the teaching of medical mycologists was researched in greater detail for this book and integrated in Chapters III and IV.

Although Emmons is widely recognized as one of the founders of medical mycology, his biography by Kwon-Chung and Campbell (1986) is a short two page obituary. His influence as a teacher and a researcher is cited as a major contribution. However, there is no mention that Emmons, through his teaching and research endeavors, proved to the medical community of his time that fungal infections were common and not rare diseases (M. McGinnis, personal communication, April 2, 1993). As a component of this study, his major contributions to the development of this discipline in the United States are examined in Chapters III and IV.

Baldwin (1981) wrote a book on the medical mycologist Elizabeth Hazen, and the chemist Rachel Brown, who were the co-discoverers of the first effective antifungal agent, nystatin. Bacon (1976) has written a biographic analysis regarding Hazen which was based upon the memoirs provided by her co-worker, Rachel Brown, Margarita Silva-Hutner, and her mentor, Gilbert Dalldorf. The impact of the Brown-Hazen grants on training and research in medical mycology is discussed in detail in Chapter IV.

Schwarz and Baum (1964) briefly described the discoveries of the most important pathogenic fungi at the end of the 1800s, first in Europe and Latin America and later in the United States. The American pioneers and their discoveries are discussed in detail in Chapter II. Robert Huntington (1985) published another brief historical account of the contributions leading to a clearer understanding of the fungus *Coccidioides immitis*. Huntington (1985) based his historical research on numerous scientific contributions that are well documented in his reference list, as well as on interviews, published scientific papers, and biographic information from key figures in medical mycology.

Chezzi (1992) wrote an editorial article on the eminent American medical mycologist, Libero Ajello, who established the first medical mycology diagnostic and research laboratory at the CDC in 1947, and related training programs later. According to Chezzi (1992), Ajello was greatly influenced by the publication of the first medical mycology textbook, *Manual of Clinical Mycology* (1945), by Norman Conant and his colleagues at Duke University. Ajello's contributions as an applied researcher and trainer are considered in more detail in Chapter III, IV and V.

Books

Ainsworth's (1986) book on the history of medical and veterinary mycology encompasses many aspects of the development of the discipline in this country, including the educational aspects. However, the focus of Ainsworth's (1986) historical work is European and does not include the last 20 or more years of development of the discipline.

Subramanian (1983) published a chronological summary of historical events in the field of mycology; however, this author focused on the discoveries involving the hyphomycetes. Although he provided comprehensive and extensive historical data, only the events related to medical mycology involving this particular group of fungi were included under Chapter I. The main thrust of his research was in general mycology.

Drouhet's (1992) historical overview of the field encompasses the training of medical mycologists in most European countries, the United States, and other areas of the world. The scope of his study did not permit an in-depth consideration of this field for any particular country or area.

Numerous obituaries and short personal sketches are found in MMSA Newsletters and other journals, some of them are listed in the references. Also other unpublished articles are found in the references.

Appendix E: List of questionnaire respondents and interviews

I.   List of questionnaire respondents (November 1993–June 1994)

94.	C Abramson, PhD		PA College of Pod. Med. at Philadelphia
57.	J Adler-Moore, PhD		Cal Poly Pomona, CA
53.	L Ajello, PhD	Retired	Centers for Disease Control at Atlanta
12.	M Alter, MS	Retired	Indiana Bd. of Health-Mycology Lab.
90.	V Anaissie, MD		Univ. of Texas, MD Anderson Cancer Cter.
6.	D Babel, PhD		Henry Ford Hospital at Detroit
92.	E Baron, PhD		
41.	K Bartizal, PhD		Merck & Co., Inc. at Rahway, NJ
73.	M Bartlett, MD		University of Indiana at Indianapolis
3.	M Belaval, MS	Retired	VA Hospital, Puerto Rico
78.	E Beneke, PhD	Retired	Michigan State University
88.	J Bennett, MD		National Institutes of Health
1.	M Berliner, PhD		Eastern Virginia Medical School at Norfolk
68.	A Berry, MD		Audie Murphy VA Hospital at San Antonio
70.	B Bolanos, PhD		University of Puerto Rico
81.	E Brummer, MD		Santa Clara Medical Center at San Jose
29.	G Bulmer, PhD	Retired	University of Oklahoma at Tulsa
27.	R Calderone, PhD		Georgetown University Medical Center
106.	A Casadevall, PhD, MD		Albert Einstein College of Medicine, New York
25.	K Clark, BS		Gen-Probe, San Diego, CA
56.	K Clemons, PhD		Santa Clara Medical Center at San Jose
87.	G Cole, PhD		University of Texas at Austin
50.	C Cooper, PhD		University of Texas Medical Branch at Galveston
64.	M Coyle, PhD		Harborview Medical Center at Seattle
84.	D Denning, MD		Monsall Hospital in England
52.	Di Salvo, MD		Nevada State Health Laboratory
67.	W Dismukes, MD		University of Alabama at Birmingham
86.	J Domer, PhD		Tulane University at New Orleans
79.	R Duma, MD	Retired	Medical College of Virginia/VCU
46.	J Feinberg, MD		Johns Hopkins University
26.	P Fidel, PhD		Harper Hospital at Detroit
55.	J Fisher, MD		Medical College of Georgia at Augusta
5.	A Folkens, PhD		Alcon Laboratories, Inc. at Fort Worth
82.	A Fothergill, BS		University of Texas Health Science Center at San Antonio
71.	R Fromtling, PhD		Merck & Co., Inc. at Rahway
100.	J Galgiani, MD		VA Medical Center at Tucson
51.	H Gallis, MD		Duke University at Durham
4.	R Gebhart, MD		VA Medical Center at Atlanta
36.	A George, MS		Merck & Co., Inc., at Rahway
96.	M Ghannoum, PhD		Harbor – UCLA Medical Center at Los Angeles
109.	N. Goodman, PhD		University of Kentucky at Lexington
14.	M Gordon, PhD	Retired	New York State Health Department at Albany

35.	J Graybill, MD		University of Texas at San Antonio
10.	A Guruswamy, BS?		Oklahoma Medical Center
76.	C Halde, PhD	Retired	University of California at San Francisco
62.	L Haley, PhD	Retired	Centers for Disease Control at Atlanta
22.	N Hall, PhD		University of Oklahoma at Oklahoma
43.	R Hamill, MD		VA Medical Center at Houston
77.	H Hardin, PhD	Retired	Duke University
24.	J Harris, PhD		Texas Department of Health
91.	R. Hopfer, PhD		University of North Carolina at Chapel Hill
97.	D Howard, PhD		University of California at Los Angeles
18.	E Jacobson, MD		VA Hospital, MCV, VCU
104.	C Jeffries, PhD		Wayne State University at Detroit
69.	W Kaplan, PhD	Retired	Centers for Disease Control at Atlanta
101.	C Kauffman, MD		VA Medical Center at Ann Arbor
11.	L Kaufman, MS, PhD		Centers for Disease Control at Atlanta
48.	T Kerkering, MD		Medical College of Virginia/VCU at Richmond
23.	T Kozel, PhD		University of Nevada at Reno
45.	C Kurtzman, PhD		National Center for Agricultural Utilization Research
32.	J Kwon-Chung, PhD		National Institutes of Health (NIH)
61.	G Land, PhD		Methodist Medical Center at Dallas
54.	D Larone, PhD		Lenox Hill Hospital at New York City
60.	R Larsen, MD		University of Southern California at Los Angeles
19.	V Layne, MD		Medical College of Virginia/VCU
47.	P Lehmann, PhD		Medical College of Ohio at Toledo
108.	H Levine, PhD	Retired	Naval Biological Laboratories, CA
8.	A Machicão, BS		University of Puerto Rico at San Juan
105.	P Magee, PhD		College of Biological Sciences at St. Paul
17.	M McGinnis, PhD		University of Texas Medical Branch at Galveston
58.	W Merz, PhD		Johns Hopkins Medical Institute at Baltimore
66.	T Mitchell, PhD		Duke University Medical Center at Durham
103.	C Morrison, PhD		Centers for Disease Control at Atlanta
102.	J Murphy, PhD		University of Oklahoma at Oklahoma City
13.	F Odds, PhD		Janssen Research Foundation
7.	G Omi, AB (Biol Sci)	Retired	
89.	D Pappagianis, MD, PhD		University of California at Davis
40.	J Perfect, MD		Duke University at Durham
20.	M Pfaller, MD		University of Iowa at Iowa City
72.	D Pincus, MS		API – Analytab Products at New Hyde
75.	E Reiss, PhD		Centers for Disease Control at Atlanta
98.	L de Repentigny, MD		STE-Justine Hospital at Montreal Quebec
93.	A Restrepo, PhD		Corp. para Investigaciones Biologicas
37.	J Rex, MD		University of Texas at Houston
38.	M Rinaldi, PhD		University of Texas Health Science Center
21.	J Rippon, PhD	Retired	University of Chicago
80.	S Rosenthal, PhD		New York Medical Center
59.	M Saag, MD		University of Alabama at Birmingham
99.	I Salkin, PhD		New York State Health Department at Albany
63.	J Shadomy, PhD	Retired	Medical College of Virginia/VCU
65.	M Silva-Hutner, PhD	Retired	Columbia University at New York
74.	J Sinski, PhD	Retired	University of Arizona at Tucson

34.	J Snyder, PhD		University of Louisville at Louisville
9.	L Steele-Moore, BS		The Medical Center of Delaware at Wilmington
28.	L Stockman, MS		Mayo Clinic at Rochester
95.	A Sugar, MD		University Hospital at Boston
85.	R Summerbell, PhD		Ontario Ministry of Health at Toronto
107.	P Szaniszlo, Ph.D		University of Texas at Austin
44.	S Thompson, MD		Emory University at Atlanta
83.	W Timberlake, PhD		
16.	C Utz, MS	Retired	
15.	J Utz, MD	Retired	Georgetown University Hospital at Washington
33.	J Vazquez, MD		The Detroit Medical Center
42.	J Vollenweider, MD		Ochsner Clinic at New Orleans
49.	I Weitzman, PhD		Columbia Presbyterian Med. Center at New York City
21.	B Wickes, PhD		University of California at San Francisco
30.	M Wieden, PhD		Veterans Affairs Medical Center at Tucson
39.	H Wilson	Retired	
31.	FT Wolf, PhD		Vanderbilt University at Nashville

Received after analysis:

| L Pasarell, MPH | University of Texas Medical Branch |
| E Koneman, MD | Denver VA Hospital at Denver |

II. Interviews From 4/1/93 to 6/27/94

L Ajello, PhD	1/4/94	Retired	Centers for Disease Control
R Beneke, PhD	2/22/94	Retired	Michigan State University
J Bennett, MD	3/17/94		National Institutes of Health
M Berliner, PhD	2/9/94		Eastern Virginia Medical School
G Bulmer, PhD	3/16/94	Retired	University of Oklahoma
C Campbell, PhD (Hon)	4/1/93	Deceased	Southern Illinois University
K Clemons, PhD	1/13/94		Santa Clara Medical Center
G Cole, PhD	3/14/94		University of Texas
G Cozad, PhD	1/19/94		University of Oklahoma
W Dismukes, MD	1/8/94		University of Alabama Medical Center
J Domer, PhD	3/15/94		Tulane University School of Medicine
N Goodman, PhD	5/24/94		University of Kentucky
M Gordon, PhD	11/18/93 & 1/10/94	Retired	New York State Health Department
C Halde, PhD	12/9/93	Retired	University of California at San Francisco
R Haley, PhD	12/31/93	Retired	Centers for Disease Control
E Jacobson, MD	1/24/94		VA Medical Center at Richmond, VA
C Jeffries, PhD	5/24/94		Wayne State University
W Kaplan, PhD	1/6/94	Retired	Centers for Disease Control
T Kerkering, MD	1/4/94		Medical College of Virginia/VCU
G Kobayashi, PhD	7/1/93, 9/1/94		Washington University
J Kwon-Chung, PhD	3/14/94 & 3/16/94		National Institutes of Health (NIH)

H Levine, PhD	6/27/94	Retired	Naval Biological Laboratories
M McGinnis, PhD	7/22/93, 7/16/94 & 9/12/94		University of Texas Medical Branch
W Merz, PhD	1/5/94		Johns Hopkins Medical Institute
T Mitchell, PhD	1/12/94		Duke University Medical Center
C Morrison, PhD	5/25/94		Centers for Disease Control
J Murphy, PhD	3/15/94		University of Oklahoma
A Padhye, PhD	3/16/94		Centers for Disease Control
E Reiss, PhD	1/25/94		Centers for Disease Control
M Rinaldi, PhD	1/3/94		University of Texas Health Science Center
J Rippon, PhD	3/14/94 & 3/16/94	Retired	University of Chicago
M Saag, MD	12/21/93		University of Alabama
I Salkin, PhD	2/22/94		New York State Health Department
J Shadomy, PhD	1/17/94	Retired	Medical College of Virginia/VCU
M Silva-Hutner, PhD	11/20/93 & 11/21/93	Retired	Columbia University
J Utz, MD	1/7/94	Retired	Georgetown University Hospital
T Walsh, MD	3/15/94		National Cancer Institute

Author Index

Subject Index

220